THE
HARDNESS OF

BY

D. TABOR
Department of Physical Chemistry
Cambridge

CLARENDON PRESS · OXFORD

OXFORD
UNIVERSITY PRESS

Great Clarendon Street, Oxford OX2 6DP

Oxford University Press is a department of the University of Oxford.
It furthers the University's objective of excellence in research, scholarship,
and education by publishing worldwide in

Oxford New York

Athens Auckland Bangkok Bogotá Buenos Aires Calcutta
Cape Town Chennai Dar es Salaam Delhi Florence Hong Kong Istanbul
Karachi Kuala Lumpur Madrid Melbourne Mexico City Mumbai
Nairobi Paris São Paulo Singapore Taipei Tokyo Toronto Warsaw

with associated companies in Berlin Ibadan

First published 1951
Published in the Oxford Classics Series 2000

British Library Cataloguing in Publication Data

Data available

Library of Congress Cataloging in Publication Data

Data available

ISBN 0 19 850776 3

1 3 5 7 9 10 8 6 4 2

Printed in Great Britain on acid-free paper by
Redwood Books Ltd., Trowbridge

PREFACE

FOR the last fifty years or more engineers and metallurgists have been making hardness measurements of metals as a means of assessing their general mechanical properties. This book is an attempt to explain hardness measurements in terms of some of the more basic physical properties of metals. It does not deal with the atomic and crystalline mechanism of elastic or plastic deformation, but starting with the assumption that metals possess certain elastic and plastic characteristics it shows that the hardness behaviour of metals may be expressed in terms of these characteristics. It is hoped that this presentation will provide, for physicists, engineers, and metallurgists, a better understanding of what hardness means and what hardness measurements imply. In writing this book a good deal of emphasis has been placed on the physical concepts involved, so that the non-mathematical reader may grasp and appreciate the general physical picture, without following the more detailed treatment.

No attempt has been made to describe in detail the practical techniques involved in hardness measurements. These have already been described in a masterly fashion in H. O'Neill's book on *The Hardness of Metals and Its Measurement* (Chapman and Hall, London, 1934), which contains a full bibliography up to 1933; in S. R. Williams's book on *Hardness and Hardness Measurements* (Amer. Soc. Met. 1942), which contains an extensive bibliography up to 1941; and in F. C. Lea's book on *Hardness of Metals* (Griffin, London, 1936). There is, in addition, a short booklet on *Hardness* by D. Landau (The Nitralloy Corporation, N.Y., 1943), and I have found a good deal of helpful information in a mimeographed article by A. F. Dunbar on *A Critical Survey of Hardness Tests* (Australian Institute of Metals, Conference 1945).

My indebtedness to others is very great: to Mr. G. Brinson of the Division of Tribophysics, C.S.I.R.O., Melbourne, who carried out some of the earlier experiments; to Dr. W. Boas and Prof. T. M. Cherry (Melbourne), to Mrs. Tipper, Dr. E. Orowan, and especially Dr. N. F. Nye and Dr. R. Hill (Cambridge) for helpful discussions; and to the Engineering Department, Cambridge University, and some of the staff, for the generous way in which they made available their testing equipment and their time.

The work described here arose out of an earlier investigation in Cambridge, between 1936 and 1939, on the area of contact between metal surfaces. The experimental work on hardness was commenced in the Division of Tribophysics, C.S.I.R.O., Melbourne, in 1945 and has been continued, amongst other work, in the Research Laboratory on the Physics and Chemistry of Surfaces, Cambridge. Throughout this period I have been very grateful to Dr. F. P. Bowden for his constant encouragement.

Except where otherwise stated, the experimental results quoted here have been obtained by the author or his colleagues.

D. T.

RESEARCH LABORATORY ON THE PHYSICS
AND CHEMISTRY OF SURFACES
DEPARTMENT OF PHYSICAL CHEMISTRY
CAMBRIDGE
January 1950

CONTENTS

ERRATA

Page 44, Section heading, *for* **Initial Plastic deformation** *read*
Initial Elastic deformation
Page 89, line 3, *for* Fig. 20, p. 42. *read* Fig. 27, p. 49.
Page 153, line 13, *for* (of the order R/n) *read* (of the order $R/\sqrt{n}$)

CHAPTER I

INTRODUCTION

In his introductory essay on the hardness of metals O'Neill (1934) has wisely observed that hardness, 'like the storminess of the seas, is easily appreciated but not readily measured'. In general hardness implies the resistance to deformation. If we accept the practical conclusion that a hard body is one that is unyielding to the touch, it is at once evident that steel is harder than rubber. If, however, we think of hardness as the ability of a body to resist permanent deformation, a substance such as rubber would appear to be harder than most metals. This is because the range over which rubber can deform elastically is very much larger than that of metals. Indeed with rubber-like materials the elastic properties play a very important part in the assessment of hardness. With metals, however, the position is different, for although the elastic moduli are large, the *range* over which metals deform elastically is relatively small. Consequently, when metals are deformed or indented (as when we attempt to estimate their hardness) the deformation is predominantly outside the elastic range and often involves considerable plastic or permanent deformation. For this reason, as we shall see, the hardness of metals is bound up primarily with their plastic properties and only to a secondary extent with their elastic properties. In some cases, however, particularly in dynamic hardness measurements, the elastic properties may be as important as the plastic properties of the metals.

Hardness measurements usually fall into three main categories: scratch hardness, indentation hardness, and rebound or dynamic hardness.

Scratch hardness

Scratch hardness is the oldest form of hardness measurement and was probably first developed by mineralogists. It depends on the ability of one solid to scratch another or to be scratched *by* another solid. The method was first put on a semi-quantitative basis by Mohs (1822), who selected ten minerals as standards, beginning with talc (scratch hardness 1) and ending with diamond (scratch hardness 10). Some typical values are given in Table I.

TABLE I

Mohs Hardness Scale

Material (Minerals)	Mohs hardness	Material (Metals)	Mohs hardness	Material (Miscellaneous)	Mohs hardness
Talc	1	Lead	1·5	Mg(OH)$_2$	1·5
Gypsum	2	Tin, cadmium	2	Finger-nail	2–2·5
Calcite	3	Aluminium	2·3–2·9	Cu$_2$O	3·5–4
Fluorite	4	Gold, Mg, Zn	2·5	ZnO	4–4·5
Apatite	5	Silver	2·7	Mn$_3$O$_4$	5–5·5
Orthoclase	6	Antimony	3	Fe$_2$O$_3$	5·5–6
Quartz	7	Copper	3	MgO	6
Topaz	8	Iron	3·5–4·5	Mn$_2$O$_3$	6·5
Corundum	9	Nickel	3·5–5	SnO$_2$	6·5–7
Diamond	10	Chromium (soft)	4·5	Martensite	7
		Cobalt	5	MoC	7–8
		Rhodium	6	V$_2$C$_3$	8
		Osmium, tantalum, tungsten, silicon, manganese	7	TiC	8–9
				Al$_2$O$_3$ (sapphire)	9
		Chromium (hard electro-deposit)	8	Mo$_2$C; SiC; VC W$_2$C; WC	9–10
		Case-hardened steel	8	Boron diamond	10+

The Mohs hardness scale has been widely used by mineralogists and lapidaries. It is not, however, well suited for metals since the intervals are not well spaced in the higher ranges of hardness and most harder metals in fact have a Mohs hardness ranging between 4 and 8. Further, the actual values obtained may depend in an unpredictable way on the experimental procedure, in particular on the inclination and orientation of the scratching edge.

Another type of scratch hardness which is a logical development of the Mohs scale consists of drawing a diamond stylus, under a definite load, across the surface to be examined. The hardness is determined by the width or depth of the resulting scratch; the harder the material the smaller the scratch (see, for example, Bierbaum, 1920; Hankins, 1923; O'Neill, 1928). The method has some value as a means of measuring the variation in hardness across a grain boundary (O'Neill, 1928). In general, however, the scratch sclerometer is a difficult instrument to operate. In addition the scratching process itself is a complicated function of the elastic, plastic, and frictional properties of the surfaces so that the method does not easily lend itself to a theoretical analysis. We shall not discuss scratch hardness any further here.

Static indentation hardness

The methods most widely used in determining the hardness of metals are static indentation methods. These involve the formation of a permanent indentation in the surface of the metal to be examined, the hardness being determined by the load and the size of the indentation formed. Because of the importance of indentation methods in hardness measurements a general discussion of the deformation and indentation of plastic materials is given in Chapter III.

In the Brinell test (Brinell, 1900; Meyer, 1908) the indenter consists of a hard steel ball, though in examining very hard metals the spherical indenter may be made of tungsten carbide or even of diamond. The general characteristics of the Brinell test are described in Chapter II and a theoretical explanation is provided in Chapters IV, V, and VI. Another type of indenter which has received wide use is the conical or pyramidal indenter as used in the Ludwik (1908) and Vickers (see Smith and Sandland, 1925) hardness tests respectively. These indenters are now usually made of diamond. The hardness behaviour is different from that observed with spherical indenters and the characteristics are described and explained in Chapter VII. Other types of indenters have at various times been described, but they are not in wide use and do not involve new principles. For this reason our discussion of static indentation tests will be restricted to spherical and pyramidal (or conical) indenters.

As we shall see, the indentation hardness of metals may in general be expressed in terms of the plastic and, to a lesser extent, the elastic properties of the metals concerned.

Dynamic hardness

Another type of hardness measurement is that involving the dynamic deformation or indentation of the metal specimen. In the most direct method an indenter is dropped on to the metal surface and the hardness is expressed in terms of the energy of impact and the size of the remaining indentation (Martel, 1895). In the Shore rebound scleroscope (Shore, 1918) the hardness is expressed in terms of the height of rebound of the indenter. These methods are discussed in Chapter VIII, and it will be shown that dynamic hardness may be expressed quantitatively in terms of the plastic and elastic properties of the metal. Another method which is, in a sense, a dynamic test is that employed in the pendulum apparatus developed by Herbert in 1923. Here an inverted compound pendulum is supported on a hard steel ball which rests on the metal under examination. The hardness is measured by the damping produced as the pendulum swings from side to side. This method is of considerable interest, but it does not lend itself readily to theoretical treatment and will not be discussed further.

Area of contact

Finally in Chapter IX we shall discuss the area of real contact between metal surfaces. This is often of considerable interest in engineering practice and the conclusions derived in the earlier chapters may be readily applied to this problem. As we shall see, the area of real contact depends little on the apparent area of the surfaces and is determined essentially by the plastic properties and, to a lesser extent, by the elastic properties of the surfaces and of the surface asperities.

REFERENCES

BIERBAUM, C. (1920), *Iron Age*, **106**, 1; (1923), *Trans. Amer. Inst. Mining Met. Engrs.* **69**, 972; (1930), *Metal Progress*, **18**, 42.

BRINELL, J. A. (1900), *II. Cong. Int. Méthodes d'Essai*, Paris. For the first English account see A. Wahlberg (1901), *J. Iron & Steel Inst.* **59, 243**.

HANKINS, G. A. (1923), *Proc. Instn. Mech. Engrs.* **1**, 423.

HERBERT, E. G. (1923), *Engineer*, **135**, 390, 686.

LUDWIK, P. (1908), *Die Kegelprobe*, J. Springer, Berlin.

MARTEL, R. (1895), *Commission des Méthodes d'Essai des Matériaux de Construction*, Paris, **3**, 261.

MEYER, E. (1908), *Zeits. d. Vercines Deutsch. Ingenieure*, **52**, 645.

MOHS, F. (1822), *Grundriss der Mineralogie*, Dresden.

O'NEILL, H. (1928), *Carnegie Schol. Memoirs, Iron & Steel Inst.* **17**, 109.

—— (1934), *The Hardness of Metals and Its Measurement*, Chapman and Hall, London.

SHORE, A. F. (1918), *J. Iron & Steel Inst.* **2**, 59. (Rebound sc.)

SMITH, R., and SANDLAND, G. (1922), *Proc. Instn. Mech. Engrs.* **1**, 623; (1925), *J. Iron & Steel Inst.* **1**, 285.

For general accounts of hardness measurements, the following books may be consulted:

LEA, F. C. (1936), *Hardness of Metals*, Charles Griffin & Co., London.

LYSAGHT, V. E. (1949), *Indentation Hardness Testing*, Reinhold. Pub. Corp. N.Y. This book appeared after the present monograph was prepared for publication. It contains a very good descriptive account of the Rockwell tests, a full account of the Knoop test, and some interesting observations on the hardness of non-metals.

O'NEILL, H. (1934), *The Hardness of Metals and Its Measurement*, Chapman and Hall, London. This contains a full bibliography up to 1933.

WILLIAMS, S. R. (1942), *Hardness and Hardness Measurements*, Amer. Soc. Met. This contains a full bibliography up to 1941.

CHAPTER II

HARDNESS MEASUREMENTS BY
SPHERICAL INDENTERS

Brinell hardness

IN the Brinell hardness test (Brinell, 1900) a hard spherical indenter is pressed under a fixed normal load on to the smooth surface of the metal under examination. When equilibrium has been reached, say after 15 or 30 sec., the load and indenter are removed and the diameter of the permanent impression measured. The Brinell hardness number (B.H.N.) is then expressed as the ratio of the load W to the *curved* area of the indentation. Hence if D is the diameter of the ball and d the chordal diameter of the indentation,

$$\text{B.H.N.} = \frac{2W}{\pi D^2[1-\sqrt{\{1-(d/D)^2\}}]}. \tag{1}$$

In most cases the B.H.N. is not a constant for a given metal but depends on the load and the size of the ball. On general physical principles we should expect that for geometrically similar indentations, whatever their actual size, the hardness number should be constant. This is found to be the case. If a ball of diameter D_1 produces an indentation of diameter d_1, the hardness number will be the same as that obtained with a ball of diameter D_2 producing an indentation of diameter d_2, provided the indentations are geometrically similar, i.e. provided the angle of indentation ϕ is the same in both cases (see Fig. 1 a). This occurs when $d_1/D_1 = d_2/D_2$.

A little consideration will show that the B.H.N. is not a satisfactory physical concept, for the ratio of the load to the curved area of the indentation does *not* give the mean pressure over the surface of the indentation. Let us assume that the mean pressure is P. If there is no friction between the surface of the indenter and the indentation the pressure is normal to the surface of the indentation. Consider the forces on an annulus of radius x and width ds (Fig. 1 b). The area of the annulus lying on the curved surface of the indentation is $2\pi x \, ds$ and the force on it is $P2\pi x \, ds$. The horizontal component of this force, by symmetry considerations, is zero. The vertical component is $P2\pi x \, dx$. If we take the sum over the whole surface area of the indentation, the resultant horizontal force, as we should

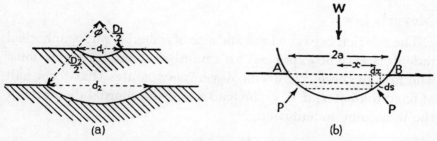

FIG. 1. (a) Geometrically similar indentations produced by spherical indenters of different diameters. (b) Calculation of mean pressure between a spherical indenter and the indentation assuming that there is no friction at the interface.

expect, is zero. The resultant vertical force which is equal to the normal load W is simply

$$W = \int_0^a P2\pi x \, dx = P\pi a^2,$$

where $2a$ is the chordal diameter of the indentation. Consequently

$$P = \frac{W}{\pi a^2}.$$

Thus the mean pressure between the surface of the indenter and the indentation is equal to the ratio of the load to the *projected* area of the indentation.† This quantity, as a measure of the hardness, was first proposed by Meyer in 1908 and is referred to as the Meyer hardness. Thus

$$\text{Meyer hardness} = \frac{4W}{\pi d^2}. \tag{2}$$

The use of the curved area in the Brinell test was originally introduced to try and compensate for the effect of work-hardening. As we shall see, however, this only introduces

† This also applies to conical and pyramidal indenters, discussed in Chapter VII. The mean pressure between the indenter and the indentation is again given by the ratio of the load to the *projected* area of the indentation.

further complications and the Meyer hardness is a much more satisfactory concept. We may note that both the Brinell and the Meyer hardness numbers have the dimensions of pressure and are usually expressed as kg./mm.[2] These may be readily converted to tons/sq. in. by dividing by 1·575.

Meyer's law

The relation between load and size of indentation for spherical indenters may be expressed by a number of empirical relations. The first of these, known as Meyer's law, states that for a ball of fixed diameter, if W is the load and d the chordal diameter of the remaining indentation,

$$W = kd^n, \tag{3}$$

where k and n are constants for the material under examination. The value of n is generally greater than 2 and usually lies between 2 and 2·5. It is found that for completely unworked materials, i.e. for fully annealed metals, n has a value near 2·5; while for fully work-hardened materials it is close to 2. This is shown in Fig. 2 for annealed and deformed copper and aluminium and for worked steel where W and d have been plotted on logarithmic ordinates. The curves are straight lines of which the slope is numerically equal to the Meyer index n, while the value of W at which d is equal to 1 is numerically equal to k. This method of analysing indentation measurements is known as the Meyer analysis since it provides a simple means of deriving both k and n.

When balls of different diameters are used, the values of k and n change. For balls of diameters $D_1, D_2, D_3,...$, giving impressions of chordal diameters $d_1, d_2, d_3,...$, a series of relations is obtained of the type

$$W = k_1 d_1^{n_1} = k_2 d_2^{n_2} = k_3 d_3^{n_3}.... \tag{4}$$

In a very extensive series of investigations Meyer (1908) found experimentally that the index n was almost independent of D but that k decreased with increasing D in such a way that

$$A = k_1 D_1^{n-2} = k_2 D_2^{n-2} = k_3 D_3^{n-2}..., \tag{5}$$

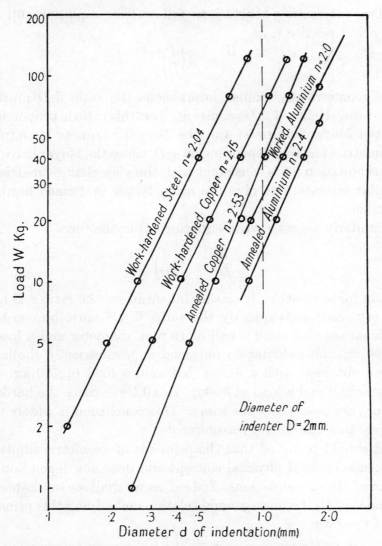

FIG. 2. Plot of load W against chordal diameter d of indentation produced by a hard spherical indenter in a metal surface; logarithmic ordinates. For annealed metals the slope n is nearly 2.5; for work-hardened metals it is nearer 2. The relation between W and d is given by $W = kd^n$, where n is Meyer's index. The value of W at which $d = 1$ is numerically equal to k.

where A is a constant. Thus the most general relation involving both d and D is

$$W = \frac{A d_1^n}{D_1^{n-2}} = \frac{A d_2^n}{D_2^{n-2}} = \frac{A d_3^n}{D_3^{n-2}} = \ldots \qquad (6)$$

Two interesting conclusions follow from equation (6). We may first rewrite it as

$$\frac{W}{d^2} = A\left(\frac{d}{D}\right)^{n-2}. \tag{6a}$$

For geometrically similar impressions the ratio d/D must be constant. Hence W/d^2 is constant. But this ratio is proportional to the Meyer hardness; and the Brinell hardness is simply a geometrical factor, depending on d/D, times the Meyer hardness. Thus equation (6) is consistent with the view that geometrically similar indentations give the same Meyer or Brinell hardness number.

Similarly we may write equation (6) in the form

$$\frac{W}{D^2} = A\left(\frac{d}{D}\right)^{n}. \tag{6b}$$

Again for geometrically similar indentations the ratio d/D must be constant; consequently the ratio W/D^2 must be constant. This means that with a ball of 10 mm. diameter and a load of 3,000 kg. the indentation obtained is geometrically similar to those obtained with a 5-mm. ball and a load of 750 kg. or a 1-mm. ball and a load of 30 kg. In all three cases the hardness values are essentially the same. This conclusion is widely used in practical hardness measurements.

It should be noted that the principle of geometric similarity is a fundamental physical concept and does not depend on the form of Meyer's equations. Indeed, as we shall see in Chapter V, equation (2), the most general relation embodying this principle is of the form

$$\frac{W}{d^2} = \psi\left(\frac{d}{D}\right), \tag{6c}$$

where ψ is some suitable function. Equation (6a) is simply a special case of equation (6c), so that the Meyer relation given in equation (6) is essentially a special case of the principle of geometric similarity rather than the explanation of it.

On the other hand, equation (6b) depends on the form of the Meyer relation. As the Meyer relation is not exact, the equivalence of a 10-mm. ball with a 3,000-kg. load and a 1-mm. ball with a 30-kg. load is not exact. For most practical purposes, however, the difference is small enough to be ignored.

In practical hardness tests the diameter of the indentation may be measured directly as in Brinell's or Meyer's method. In other tests the depth of the indentation may be measured as in the Rockwell test, or the load required to give a fixed depth of indentation may be determined as in the Monotron test. These and other practical methods are described in O'Neill, but will only be discussed briefly in this book (see Chapter VII). It is clear that the fundamental behaviour is determined by the two empirical equations of Meyer. In the following chapters we shall provide a theoretical basis for these equations and show how the hardness may be expressed in terms of more fundamental physical properties of the metals. Before doing so, however, it is of some interest to consider a few practical points connected with hardness tests involving the use of spherical indenters.

Comparison of Brinell and Meyer hardness

It is interesting to compare the Brinell and the Meyer hardness values at various loads. The results for annealed and highly deformed copper specimens are plotted in Fig. 3. It is seen that, for the highly worked copper, the Meyer hardness is essentially constant and independent of the load, i.e. the mean pressure resisting indentation is approximately constant, corresponding to a value of 2 for the Meyer index n. The B.H.N., however, at first nearly constant, falls with increasing load because of the increase in size of the curved area of the indentation. Thus the B.H.N. at higher loads may be smaller than at smaller loads, giving the impression that the metal is softer for larger indentations.

This misleading characteristic of the B.H.N. is also marked with materials which can undergo appreciable work-hardening. For example, with annealed copper the mean pressure resisting indentation, that is the Meyer hardness, increases steadily as the load (and hence the size of the indentation) is increased. This is attributed to the work-hardening of the metal by the indentation process itself. The B.H.N., however, again rises at first, and then falls with increasing load. Thus the B.H.N. at high loads might lead us to believe that the metal has been less work-hardened than at medium loads. This is clearly not the case.

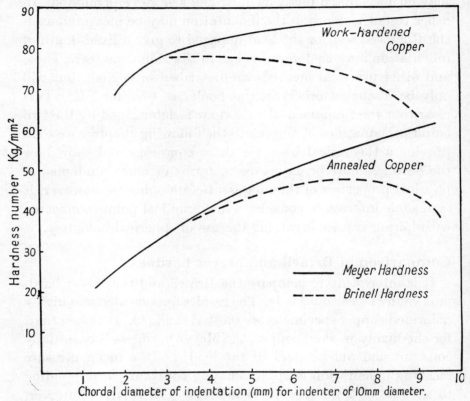

FIG. 3. Brinell hardness and Meyer hardness values for annealed and work-hardened copper as the size of the indentation is increased. The Meyer hardness values lie on a monotonic curve. The Brinell hardness values first increase and then decrease for large indentations as a result of the increasing area of the curved surface of the indentation. The Brinell values thus make it appear that for large indentations the metal is softer than for small indentations.

It is evident that the Meyer hardness number is a more satis-factory and more fundamental concept in the measurement of indentation hardness.

Validity of Meyer's law

Meyer's relation $W = kd^n$ is found to hold for balls sub-merged almost to their diameters. There is, however, a lower limit to the validity of Meyer's law. For small loads, giving small indentations, logarithmic plotting of the points gives

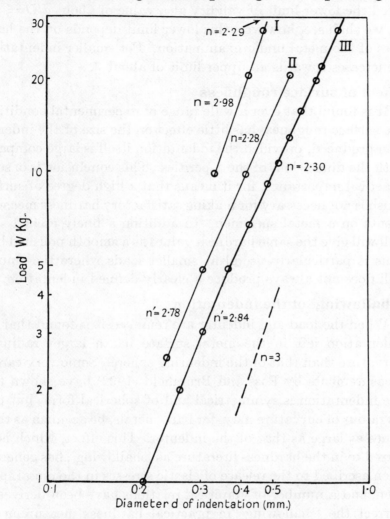

FIG. 4. Plot of W against d. The relation $W = kd^n$ is valid for larger indentations but for smaller indentations n increases towards an upper value of about 3. Curve I, steel W, $D = 10$ mm.: Curve II, steel A, $D = 10$ mm.: Curve III, soft iron, $D = 20$ mm. (From data given by O'Neill, 1934.)

unduly high values of the index n. This effect is shown in some results taken from O'Neill's book for balls of diameter $D = 10$ mm. and $D = 20$ mm. (see Fig. 4). It is seen that for indentations greater than about $d = 0.5$ mm., Meyer's law is obeyed and the normal values of n are obtained. Meyer, in fact, fixed the lower limit of validity at a value of about $d/D = 0.1$. As we shall see, however, this lower limit depends on the hardness of the metal under examination. For smaller indentations n increases towards an upper limit of about 3.

Effect of surface roughness

It is found that over a wide range of experimental conditions the surface roughness has little effect on the size of the indentation produced, provided the indentation itself is large compared with the dimensions of the asperities. This conclusion is of some practical importance, for it means that a high degree of surface finish is not necessary for making satisfactory hardness measurements on a metal specimen. In addition a finely etched steel ball will give the same hardness values as a smooth polished ball. This is particularly useful at smaller loads where the smooth ball does not always produce a clearly defined indentation.

'Shallowing' of the indentation

When the load and indenter are removed it is found that the indentation left in the metal surface has a larger radius of curvature than that of the indenting sphere. Some very careful measurements by Foss and Brumfield (1922) have shown that the indentation is symmetrical and of spherical form, but that its radius of curvature may, for hard metals, be as much as three times as large as that of the indenter. This effect, which is referred to in the hardness literature as 'shallowing', has generally been ascribed to the release of elastic stresses in the metal specimen, and a number of empirical relations have been derived to correct the 'shallowing' in practical hardness measurements. Little work, however, of an analytical nature has been carried out to relate this directly to the elastic properties of the metal and the indenter.

There is also a diminution in the chordal diameter of the indentation when the indenter is removed, but in general the effect is small and does not as a rule exceed a few per cent. It follows that, because of shallowing, depth measurements of the 'recovered' indentation are much more unreliable than measurements of the chordal diameter of the 'recovered' indentation. In effect the mean pressure derived from the diameter of the 'recovered' indentation will be the same, to within a few per cent., as the mean pressure existing between the indenter and the surface before the load is removed.

'Piling-up' and 'sinking-in'

As a result of the displacement of metal from the indentation itself there is appreciable deformation of metal around the

(a)   (b)

FIG. 5. Deformation around the indentation produced by a spherical indenter: (a) 'piling-up' which is observed with highly worked materials, (b) 'sinking-in' which is observed with annealed materials. The effects have been exaggerated to show more clearly the deformation relative to the original level (dotted line).

indentation. In some cases there is an upward extrusion of displaced metal so as to form a raised crater. The overall diameter of this displaced metal on the surface may be twice the diameter of the indentation and the effect is known as 'piling-up' (Fig. 5 a). This effect is most marked with materials which are highly work-hardened, for which the Meyer index n is nearer 2. On the other hand, with annealed metals, there is a tendency for the metal to be depressed around the indentation and this effect is known as 'sinking-in' (Fig. 5 b). This depression is only observed close to the rim of the indentation, and at distances well removed from the indentation a slight elevation of the metal above the original surface level is usually found. Both the 'piling-up' and the 'sinking-in' contribute some uncertainty to the diameter of the indentation and a number of empirical corrections for these effects have been proposed.

'Strainless' indentation

It has long been recognized that the formation of the indentation itself leads to an increase in the effective hardness of the metal so that the hardness test itself changes the hardness of the metal under examination. This work-hardening of the metal during the indentation process is accounted for by the empirical relations of Meyer. A number of attempts have, however, been made to determine the 'absolute' hardness of metals by a method which does not produce any work-hardening. This can only be achieved if the method does not produce appreciable plastic deformation of the metal, i.e. if it is plastically strainless. Two 'strainless' methods have been attempted. The first, due to Harris (1922), consists of making a series of successive impressions on the same spot with a fixed load, the strain-hardening being removed by annealing between each application of the load. After about 10 anneals Harris found that the load finally 'floated' on the impression and produced no further increase in the size of the indentation. At this stage, therefore, the indenter was supported solely by the elastic stresses in the metal. Starting with annealed copper which gives an initial Meyer hardness of about 40 kg./mm.^{2}, Harris found that the 'absolute' hardness at the end of the above series of indentations had fallen to about 15 kg./mm.^{2} This is close to the value obtained for direct indentation measurements at very small loads.

Harris's method is limited to annealed metals. Mahin and Foss (1939) prepared cavities in the metal under examination with a hemispherical cutting-tool possessing a cutting radius of 5 mm. The cavities were finished by hand grinding and polishing, and it was assumed that this method of preparing the cavity had not produced appreciable work-hardening of the metal. Using an indenter of diameter 10 mm., experiments were then carried out to determine which cavity could support a given load without suffering enlargement, i.e. without undergoing plastic deformation. The Meyer hardness at this stage was considered to be the 'absolute' hardness of the metal, and these workers again found that the 'absolute' hardness was about one-third that of the 'normal' hardness of the metal.

Ultimate tensile strength and B.H.N.

One of the reasons for the wide use of the Brinell test in engineering is the existence of an empirical relation between the

B.H.N. and the ultimate tensile strength of a metal. The latter is the maximum nominal stress that a metal can experience in tension before it fails and is the ratio of the maximum load to the *original* cross-section of the tensile specimen. This quantity is of considerable practical importance, and the ratio connecting it with the B.H.N. is given in the following table, which is taken from data given by Greaves and Jones (1926). The B.H.N. is in kg./mm.^{2}

TABLE II

Metal	Ratio of $\dfrac{Ultimate\ tensile\ strength}{B.H.N.}$	
	kg./mm.^{2}	tons/in.^{2}
Heat-treated alloy steels .	0·33	0·21
Heat-treated carbon steels .	0·34	0·215
Medium carbon steels . .	0·35	0·22
Mild steel	0·36	0·23

It is seen that for the steels given in this table the maximum nominal stress (or ultimate tensile strength) is approximately one-third the B.H.N. when expressed in kg./mm.^{2} The same ratio is found to hold approximately for most metals which do not work-harden appreciably. On the other hand, for metals which are capable of appreciable work-hardening the ratio is considerably larger. For example, for annealed nickel it is about 0·49, for certain austenitic steels about 0·52, and for annealed copper about 0·55. Summarizing these results we may quote O'Neill's conclusion that, although there is no constant ratio for *all* metals, if the Meyer index $n = 2 \cdot 2$ or less, the ratio is approximately 0·36 (for kg./mm.^{2}). If the value of the Meyer index n is higher than 2·2 the ratio is usually greater than this.

In Chapters IV, V, and VI we shall show that Meyer's equations and the points described above may be explained in terms of the general physical processes involved when a spherical indenter penetrates a plastic material. Before dealing with this, we shall first discuss the more general problem of the deformation and indentation of plastic materials.

REFERENCES

BRINELL, J. A. (1900), *II. Cong. Int. Méthodes d'Essai*, Paris. For the first English account see A. Wahlberg (1901), *J. Iron & Steel Inst.* **59**, 243.

FOSS, F., and BRUMFIELD, R. (1922), *Proc. Amer. Soc. Test. Mat.* **22**, 312.

GREAVES, R., and JONES, J. A. (1926), *J. Iron & Steel Inst.* **1**, 335.

HARRIS, F. W. (1922), *J. Inst. Metals*, **28**, 327.

MAHIN, E. G., and FOSS, G. J. (1939), *Trans. A.S.M.* **27**, 337.

MEYER, E. (1908), *Zeits. d. Vereines Deutsch. Ingenieure*, **52**, 645, 740, 835.

O'NEILL, H. (1934), *The Hardness of Metals and Its Measurement*, Chapman and Hall, London.

CHAPTER III

THE DEFORMATION AND INDENTATION
OF IDEAL PLASTIC METALS

IN the formation of the indentation made in Brinell hardness measurements we are concerned mainly with the plastic flow of the metal around the indenter. For this reason we shall first discuss briefly the nature of plastic deformation.

Stress and strain

Suppose we take a uniform cylinder of metal and subject it to a tensile or compressive stress along its axis. We measure the change in dimensions, i.e. the strain as a function of the deforming stress. The stress may be defined in two ways. If the cylinder maintains a uniform section as it is extended or compressed, the *true* stress is the tensile or compressive force divided by the cross-sectional area of the specimen at that stage of the deformation. On the other hand, the *nominal* stress is the tensile or compressive force divided by the original cross-sectional area of the specimen.

The strain may also be defined in several ways. The simplest, known as the linear strain, is the fractional increase in length of the specimen, so that if l_0 is the original length and l the length at any later stage, $\epsilon = (l-l_0)/l_0$. For extension ϵ is positive, for compression it is negative. Instead of using the lengths we may use the cross-sectional areas A_0 and A. Since in plastic deformation there is a negligible volume change, $l_0 A_0 = lA = $ constant, so that for plastic deformation

$$\epsilon = (l-l_0)/l_0 = \frac{A_0 l - Al}{Al} = \frac{A_0 - A}{A}.$$

Another way of measuring strain (which has been widely used in wire-drawing) is in terms of the reduction of area. We may call this the areal strain, and it is defined as

$$\epsilon_r = \frac{A_0 - A}{A_0}.$$

It is seen that this differs somewhat from the linear strain ϵ, especially for large reductions. We may also note that it is positive for extension and negative for compression.

If large deformations are involved it is often more satisfactory to use the logarithmic strain ϵ^*. This is obtained from the element of strain at any stage $d\epsilon^* = (dl)/l$ so that $\epsilon^* = \log_e l/l_0$. One case in which the advantage of the logarithmic strain is apparent is in the comparison of extension and compression experiments. If a uniform cylinder is extended uniformly to twice its length, the linear strain is $\epsilon = (2l_0 - l_0)/l_0 = 100$ per cent. In compression the specimen would have to be squeezed to zero thickness to give a linear (negative) strain of 100 per cent. On general grounds we should expect the strain on compressing a cylinder to half its length to be the same (though opposite in sign) as that obtained on extending the cylinder to twice its length. This occurs with logarithmic strain, since for extension $\epsilon^* = \log_e 2$ whilst for compression

$$\epsilon^* = \log_e \tfrac{1}{2} = \log_e 1 - \log_e 2 = -\log_e 2.$$

It is interesting to note that the same agreement between tensile and compressive strain is obtained if we use the linear strain ϵ in the tensile experiment and the areal strain ϵ_r in the compressive experiment. Thus an extension of the specimen to double its length gives a value of $\epsilon = 100$ per cent., whilst a compression to half its length gives a (negative) value of $\epsilon_r = 100$ per cent. The reason for this agreement, when the strains at first sight appear to be mutually inconsistent, is simple. According to the definition of logarithmic strain, the strain in tension is equivalent to the strain in compression when the ratio l/l_0 in tension is equal to the ratio l_0/l in compression. This equivalence remains equally valid if we subtract unity from each ratio. Thus the tensile and compressive strains are equivalent if the term $l/l_0 - 1 \; \{= (l - l_0)/l_0\}$ in tension is equal to the term $l_0/l - 1 \; \{= (l_0 - l)/l\}$ in compression. The first term is simply the linear strain ϵ. The second term, remembering that $A_0 l_0 = Al$, is the areal strain ϵ_r with a negative sign. It follows that if the logarithmic strain gives equivalent strains for tension and compression, similar agreement will be found between the linear strain for tension and the areal strain for compression.

For most purposes it is sufficient to use the linear strain, and this we shall do in the following discussion of the stress–strain characteristics of metals under tension.

True stress–strain curves under tension

We first consider the behaviour of an 'ideal' plastic metal under tension and plot the linear strain ϵ as a function of the

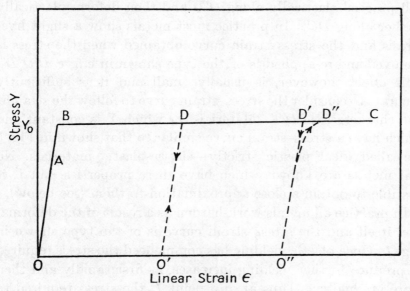

Fig. 6. True stress–strain curve for ideally plastic metal under tension. OA is the (reversible) elastic region and the slope of OA is equal to Young's modulus. The metal yields plastically at the elastic limit Y_0 (point B) and the yield stress remains constant after further deformation. At any subsequent stage, if the stress is removed, the stress–strain curve follows the reversible path DO'(or $D'O''D''$).

true stress Y. At first there will be a slight increase in length, which is proportional to the applied stress. Over this range, where Hooke's law is obeyed the deformation is elastic and on removing the tension the cylinder will return to its original length (OA, Fig. 6). The slope of the line OA, i.e. the ratio of stress to strain, is Young's modulus of the metal. When the stress reaches a certain value the cylinder will increase in length in a non-reversible way, and the stress at which this occurs is called the elastic limit Y_0. If the material does not work-harden, i.e. if the stress remains constant during extension, the stress–strain curve is a straight line (BC, Fig. 6) parallel to the strain axis. If, at any point D, the stress is reduced, the cylinder contracts elastically along the curve DO' where DO' is approximately parallel to AO. When the stress is completely removed the cylinder will have suffered a permanent plastic deformation of amount OO'. On increasing the stress again the deformation will proceed elastically along $O'D$ and then deform plastically farther along DC. In practice most metals show a slight hysteresis and the stress–strain curve obtained when the stress is removed and re-applied is of the type shown in curve $D'O''D''$. This effect, however, is usually small and it is sufficiently accurate to consider the stress–strain curve to follow the straight line characteristic DO'. Materials for which Y is constant and which have a stress–strain curve similar to that shown in Fig. 6 are called 'ideal' plastic (strictly—elastic-plastic) materials. No real metals are known which have these properties, but it is possible to obtain a close approximation to them (see below).

In practice all metals work-harden as a result of the deformation itself and the stress–strain curve is of the type shown in Fig. 7. Once plastic yielding has commenced the stress required to produce further yielding increases, at first rapidly and then more gradually.† Thus at any point D the stress required to produce further plastic flow is no longer the initial yield stress Y_0 but a larger stress, so that the yield stress varies with the amount of deformation to which the metal has been subjected. With many metals the dependence of the yield stress Y on the deformation ϵ may be expressed approximately by a relation of the type $Y = b\epsilon^x$ (Nadai, 1931). If at the point D (Fig. 7) the stress is removed, the specimen contracts elastically along the line DO'. Consequently if we use a specimen that has already

† We do not consider here the special behaviour of annealed mild steel and some aluminium alloys which show an initial yield value (the upper yield point) which is followed by a drop in the yield stress for subsequent small strains. For larger strains, however, the yield stress again increases with strain and follows the course of the curve shown in Fig. 7

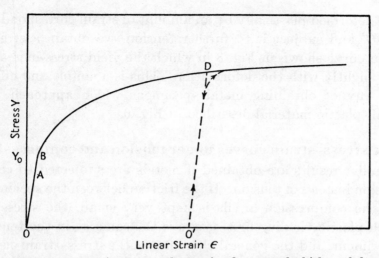

FIG. 7. True stress–strain curve under tension for a metal which work-hardens as a result of deformation. OA is the elastic region, and B marks the elastic limit or yield stress Y_0 at which plastic deformation commences. As deformation proceeds there is a steady increase in the stress at which plastic deformation occurs. At the point D if the stress is removed, the stress–strain curve follows the reversible path DO'.

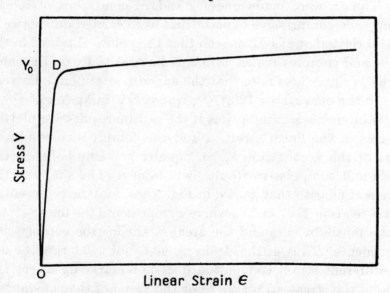

FIG. 8. True stress–strain curve of a metal which has been plastically deformed by the amount OO' in Fig. 7. The curve is similar to Fig. 7 except that the origin has been moved to O'. OD is the elastic region, D the initial yield stress at which plastic deformation first occurs. It is seen that the yield stress does not increase rapidly with further deformation so that the material is close to an ideally plastic metal (Fig. 6).

been deformed plastically by an amount of strain corresponding to OO' and subject it to further tension, we obtain a stress–strain curve shown in Fig. 8 in which the yield stress increases only slightly with the deformation. This is a simple and effective way of obtaining metal specimens which approach the 'ideal' plastic material described in Fig. 6.

True stress–strain curves under tension and compression

Similar results are obtained if metals are subjected to compression instead of tension. If the friction between the specimen and the compression anvils is kept very small, the stress at which plastic yielding first occurs is the same as in the tensile experiment, and the general character of the stress–strain curve is similar in both cases (Ludwik and Scheu, 1925). A typical example for partially annealed aluminium is shown in Fig. 9, in which the logarithmic strain ϵ^* has been plotted against the true stress Y. The tensile measurements, kindly carried out by Mrs. Tipper, were made under standard conditions of tensile testing; the compressive experiments were carried out between well-lubricated anvils. It is seen that the points obtained in the tensile and compressive experiments lie essentially on the same curve. We may also note that the experimental points lie very close to the curve $Y = 12(\epsilon^*)^{0.28}$, where Y is in kg./mm.²

It is interesting to note that if the tensile results are plotted in terms of the linear strain ϵ, and the compressive results in terms of the areal strain ϵ_r, an equally good fit between the tensile and compressive results is obtained. The shape of the curve is similar to that shown in Fig. 9 and may be represented by the relation $Y = 11\epsilon_d^{0.25}$, where ϵ_d represents the linear strain for the tensile results and the areal strain for the compressive experiments. Thus, if the strain range is not too large, the use of a different strain convention does not markedly affect the shape of the stress–strain curve or the value of the constants b and x in the relation $Y = b\epsilon^x$. In particular the equivalence of the linear strain in tension and the areal strain in compression means that the true stress–strain curve in tension may be deduced, with good accuracy, from 'frictionless' compression

experiments and vice versa. This conclusion is implicit in a
number of points discussed at the end of Chapter V.

If there is appreciable friction between the face of the specimen and the compression anvils, the stress required to produce
plastic flow at any stage of the experiment is higher than the

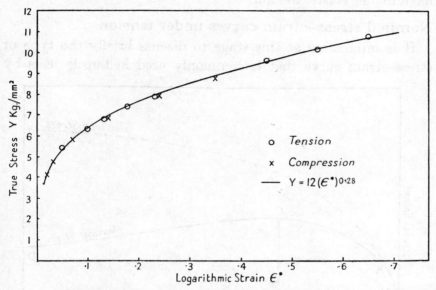

FIG. 9. True stress–strain curve ($Y-\epsilon^*$ curve) for partially annealed aluminium. o tensile experiment, ✗ compressive experiment. The full line follows
the empirical relation $Y = 12(\epsilon^*)^{0.28}$, where Y is in kg./mm.² and ϵ^* is the
logarithmic strain.

true yield stress. Under such conditions the agreement between
the tensile and compressive stress–strain curves may be poor.

There is one further point in the deformation of metals that
is worth recalling at this stage. The elastic properties of metals
are due to the interatomic forces and they are very little affected
by the amount of work-hardening to which the metal is subjected. Consequently Young's modulus, for example, depends
little on the degree of work-hardening of the metal. For this
reason the slope of the line $O'D$ in Figs. 6 and 7 is essentially
the same as that of the line OA. On the other hand, the plastic
properties depend on the slipping of the crystals along certain
crystallographic planes. Although ultimately these are also

related to the crystalline structure and the interatomic forces, there is no simple and obvious connexion between the elastic moduli and the elastic limit of metals.† It is, however, true that in general softer metals (i.e. metals with a lower elastic limit) have smaller elastic moduli, whilst harder metals usually have higher elastic moduli.

Nominal stress–strain curves under tension

It is interesting at this stage to discuss briefly the type of stress–strain curve that is commonly used in tensile tests by

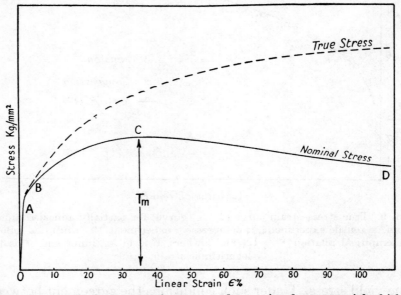

Fig. 10. Nominal stress–strain curve under tension for a material which work-hardens. The nominal stress reaches a maximum T_m after which the specimen begins to fail, but the true stress still increases.

engineers. It is usual to use the fractional increase in length for the strain, i.e. the linear strain, but to plot the nominal instead of the true stress, i.e. the load divided by the original cross-sectional area. A typical stress–strain curve of this type for a metal which work-hardens is shown in Fig. 10. The portion OA

† An interesting derivation of the yield stress of highly worked metals in terms of the atomic spacing, the crystal size, and the elastic modulus of rigidity has been proposed by Bragg (1942). Similar results may also be derived from the earlier work of G. I. Taylor (1934).

represents the initial elastic region and AB marks the early stages of plastic deformation. As the extension proceeds the cross-section of the specimen is reduced whilst the material continuously work-hardens. A point C is reached at which the nominal stress reaches its maximum value T_m, and beyond this stage the decrease in cross-section becomes more rapid than the

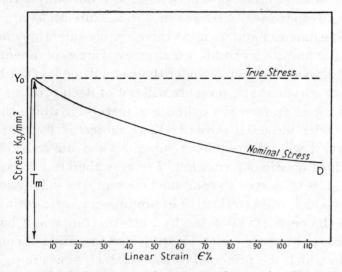

FIG. 11. Nominal stress–strain curve under tension for an ideal plastic material. As soon as plastic deformation occurs the material begins to fail so that the maximum tensile stress is very close to the initial yield stress Y_0.

increase in true yield stress and the specimen begins to fail. It finally breaks at the point D. The true stress is shown in the broken curve, and it is seen that it increases steadily the greater the amount of deformation.

Fig. 11 shows the stress–strain curve for an ideal plastic metal. As soon as plastic yielding occurs at the yield stress Y_0 the cross-section of the tensile specimen decreases, so that whilst the true yield stress remains constant the nominal yield stress steadily decreases until fracture occurs at D. It is clear that in this case the maximum nominal stress T_m is essentially the same as the yield stress Y_0. The true stress curve is again shown in the broken curve. It is evident that the true stress is a more fundamental concept than the nominal stress. In engineering practice, however, the nominal stress is a very useful and convenient quantity.

Plastic deformation under combined stresses

So far we have described the plastic yielding of a metal under a uniform tension or compression. When an indenter is pressed on to a surface, however, the stresses are not simple tensile or compressive stresses. Stresses in various directions are set up under the indenter and we must therefore consider the conditions for plastic flow under combined stresses. One experimental clue to the behaviour of metals under these conditions is the observation that a hydrostatic pressure will not of itself produce plastic deformation. In fact, if a cylinder of metal, for which the yield stress under uni-axial stress is Y, is subjected to hydrostatic pressure, it will still require a superimposed uni-axial stress Y to produce plastic deformation. We may therefore expect that if a metal is subjected to combined stresses, the only part of the stresses which will be effective in producing plastic deformation will be the part left after the hydrostatic component has been subtracted. Suppose a metal is subjected to a state of combined stress in which the principal (orthogonal) stresses are p_1, p_2, p_3. We may consider that this is equivalent to the sum of a hydrostatic component $\frac{1}{3}(p_1+p_2+p_3)$ and reduced stresses

$$p_1-\tfrac{1}{3}(p_1+p_2+p_3), \quad p_2-\tfrac{1}{3}(p_1+p_2+p_3), \quad \text{and} \quad p_3-\tfrac{1}{3}(p_1+p_2+p_3).$$

The hydrostatic component plays no part in the plastic deformation of the metal. Only the reduced stresses are responsible for producing plastic deformation, and it is found *experimentally* that this occurs when

$$[p_1-\tfrac{1}{3}(p_1+p_2+p_3)]^2+[p_2-\tfrac{1}{3}(p_1+p_2+p_3)]^2+$$
$$+[p_3-\tfrac{1}{3}(p_1+p_2+p_3)]^2 = \text{constant}$$

or $\qquad \tfrac{1}{3}[(p_1-p_2)^2+(p_2-p_3)^2+(p_3-p_1)^2] = \text{constant}.$ \hfill (1)

For uni-axial tension (or compression) $p_2 = p_3 = 0$, and plastic deformation occurs when the axial stress equals the yield stress Y, i.e. when $p_1 = Y$. This makes the value of the constant equal $\frac{2}{3}Y^2$, so that equation (1) becomes

$$(p_1-p_2)^2+(p_2-p_3)^2+(p_3-p_1)^2 = 2Y^2. \qquad (2)$$

This expression was originally derived by Huber (1904) and independently by von Mises (1913) from more general considerations of the conditions for plastic deformation. The relation is known as the Huber–Mises criterion of plasticity and it has been supported by a very large body of experimental data (see Nadai, 1931).

An alternative and much earlier criterion for plastic flow was that proposed by Tresca (1864) and later revived by Mohr (1900). This was based on the view that plastic deformation occurs when the maximum shear stress reaches a certain critical value. If p_1, p_2, p_3 are the principal stresses, the shear stresses are $\frac{1}{2}(p_1-p_2)$, $\frac{1}{2}(p_2-p_3)$, and $\frac{1}{2}(p_3-p_1)$. If $p_1 > p_2 > p_3$ the maximum shear stress is $\frac{1}{2}(p_1-p_3)$, and it was considered that this parameter determines the occurrence of plastic deformation. We may determine the actual value of this parameter by considering the special case of uni-axial tension. Here $p_2 = p_3 = 0$, so that the maximum shear stress is $\frac{1}{2}p$, and this clearly produces plastic deformation when it is equal to $\frac{1}{2}Y$. Tresca suggested that plastic flow in general occurs when the maximum shear becomes equal to $\frac{1}{2}Y$. Thus the Tresca or Mohr criterion for plastic flow is that

$$p_1-p_3 = Y \quad \text{when} \quad p_1 > p_2 > p_3. \tag{3}$$

With materials such as annealed mild steel there is evidence that the condition for plastic yielding at the upper yield point is nearer the Tresca (or Mohr) criterion than the Huber–Mises criterion. We may note, however, that there are two conditions under which these criteria become essentially the same. The first occurs when any two of the principal stresses are equal. If, for example, we substitute $p_2 = p_3$ in equation (2), the result is identical with equation (3). The second occurs in the case of two-dimensional deformation or plane strain. It may be shown that the condition for zero plastic deformation in the direction of p_2 is that $p_2 = \frac{1}{2}(p_1+p_3)$. Substituting this value for p_2 in equation (2) the Huber–Mises criterion becomes

$$p_1-p_3 = (2/\sqrt{3})Y.$$

Thus the condition for plasticity is a maximum shear-stress

condition as in equation (3), but the value of the constant is about 1·15 times larger.

Several other conditions for plastic flow under combined stresses have been suggested, but these are mainly of historical interest. One that we may mention in passing is that proposed by Haar and von Karman in 1909. They suggested that when plastic flow occurs, two of the principal stresses (say p_2 and p_3) are equal so that the metal is subjected to a hydrostatic pressure equal to p_2 (or p_3) and to a residual uni-axial stress equal to p_1-p_2 (or p_1-p_3). They suggested therefore as a criterion for plasticity

$$p_1-p_2 = Y \quad \text{and} \quad p_2 = p_3. \tag{4}$$

There is no physical basis for this assumption that two of the principal stresses are equal, but it has recently been revived by Russian workers, who find that by using this hypothesis it is possible to solve certain problems in plasticity which are otherwise intractable.

Conditions for two-dimensional plastic flow

It is outside the scope of this book to describe in detail the mathematical treatment of the problem of plastic deformation. We may, however, discuss briefly the physical principles involved and indicate the way in which these have been applied to a number of practical problems. The starting-point is the assumption that plastic flow occurs when the Huber–Mises criterion is satisfied. In two-dimensional flow this occurs when the maximum shear stress reaches a critical value k, where $2k = 1·15Y$. This is similar to the condition obtained with the Tresca–Mohr criterion, except that here $k = \frac{1}{2}Y$ or $2k = Y$. Since most of the problems solved by this method involve two-dimensional plastic deformation, it is evident that the treatment is the same whatever the criterion of plasticity adopted: there is merely a change in the value of k according as to whether the Tresca or the Huber–Mises criterion is adopted.

In a region in which the plastic strains are large compared with the elastic, we may consider the deformation to be determined primarily by the plastic properties of the metal. We shall first show that at any point where plastic flow occurs the total stresses can be represented by the sum of a hydrostatic pressure p and a shear stress k where k has the value given above. We consider any two-dimensional element which is subjected to principal (orthogonal) compressive stresses P and Q (Fig. 12).†

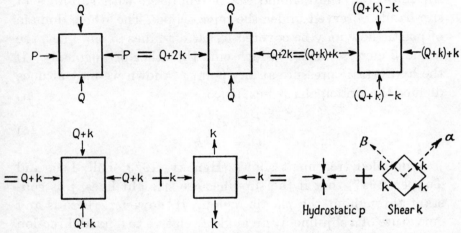

FIG. 12. Diagram showing that for two-dimensional stress, the principal stresses P and Q may be replaced by a hydrostatic pressure p and a maximum shear stress s, where $p = Q+k = P-k = \frac{1}{2}(P+Q)$ and $s = k$.

These stresses completely define the stress condition on the element. The maximum shear stress produced by these stresses is at 45° to P and Q and has a magnitude $\frac{1}{2}(P-Q)$. We know that if the material in this element is flowing plastically the maximum shear stress has reached a critical value which we put equal to k. As we saw above, $2k = Y$ according to the Tresca–Mohr criterion or $2k = 1{\cdot}15Y$ according to the Huber–Mises criterion. It follows that $P = Q+2k$. We may thus replace the principal stresses by Q and $Q+2k$. These may be written respectively as $(Q+k)-k$ and $(Q+k)+k$. The first term in each stress constitutes a hydrostatic pressure $p = Q+k$. The second terms, involving a compression k and a tension k, are equivalent to a pair of orthogonal shear stresses k at 45° to the direction

† In the usual treatment it is customary to consider *tensile* stresses as positive. The use of compressive stresses adopted here does not substantially affect the general analysis.

of P and Q. These transformations are shown diagrammatically in Fig. 12. Thus when plastic deformation occurs, the stresses at any point may be expressed in terms of a shear stress k, plus a hydrostatic pressure p; k is constant, whilst p may vary from point to point. The lines of maximum shear stress k are called slip-lines, but they should *not* be confused with slip-lines or slip-bands observed under the microscope. The whole domain of plastic flow may be covered by two families of slip-lines, the α and β curves, each of which cuts the other orthogonally. If the hydrostatic pressure at any point is known we may at once deduce the principal stresses, since

$$P = p+k,$$
$$Q = p-k. \tag{5}$$

A detailed treatment shows (Hencky, 1923; Hill, Lee, and Tupper, 1947) that if the slip-lines are straight lines, p is constant throughout the plastic region. If, however, there is any curvature of a slip-line by an angle ϕ relative to a fixed direction (ϕ positive when moving anticlockwise), then:

$$p+2k\phi = \text{constant along an } \alpha \text{ line,}$$
$$p-2k\phi = \text{constant along a } \beta \text{ line.} \tag{6}$$

This was first derived by Hencky (1923) from considerations of the equilibrium of an element. The value of the constant is determined by the boundary conditions.

We may now consider a number of simple applications of the slip-line theory. Suppose a metal is deformed over a region AB and as a result plastic flow also occurs in the material under the free surface BC (Fig. 13a). Then the slip-lines at the free surface must make an angle of 45° with the surface. This follows because there must be no resultant tangential force at the free surface. But there must also be no resultant normal stress at the surface, i.e. $Q = 0$. Hence $Q = p-k = 0$ or $p = k$ and $P = p+k = 2k$. This means that at the free surface there is a hydrostatic pressure $p = k$, and a maximum shear stress k. In terms of the principal stresses we may say that the normal stress $Q = 0$, whilst the transverse compressive stress $P = 2k$.

Similarly, if a hard indenter deforms a softer surface, the slip-lines must make an angle of 45° with the common surface of contact if there is no frictional force, since under these conditions there can be no resultant tangential force at the interface (Fig. 13 b). If, however, there is a finite frictional force, the

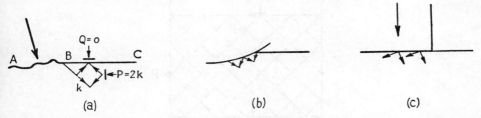

(a)   (b)   (c)

FIG. 13. Slip-lines at the surface of plastically deformed metal: (a) a free surface, (b) a frictionless interface, (c) an interface with friction. For (a) and (b) the slip-lines make an angle of 45° to the surface.

slip-lines no longer make an angle of 45° and the resultant tangential force arising from the components of the shear stress parallel to the surface is equal to the frictional force at the interface (Fig. 13 c).

Suppose a two-dimensional slab of an ideal plastic material is compressed between two flat parallel anvils (Fig. 14). We consider the plane strain problem in which the material is confined so that there is no flow normal to the plane of the paper. The anvils are considered undeformable and the friction at the interface is neglected. Then the slip-lines are at 45° to the free faces and at 45° to the anvil faces. It may be shown that the families of slip-lines are straight lines as indicated in Fig. 14, so that p is constant throughout the material. At the free surface, as we saw above, $Q = p - k = 0$, so that $p = k$. But the second orthogonal stress P is given by $P = p + k = 2k$. Hence the normal pressure on the anvil is given by $P = 2k = Y$ according to the Tresca criterion. This is, of course, what we expect, since the stress to produce plastic flow under compression is simply equal to Y. It may be noted that according to the Huber–Mises criterion the pressure is equal to $1 \cdot 15Y$. If, however, the metal were allowed to flow in the third dimension, the Huber–Mises criterion would also give a value of Y.

The effect of friction at the anvil face is too complicated to be dealt with here by the slip-line treatment. In a subsequent section we shall show that it can be treated by a simpler approximate method. The results show that the pressure is not uniform

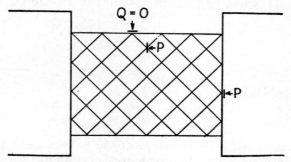

FIG. 14. Two-dimensional deformation (plane strain) of an ideal plastic metal between flat rigid anvils. The slip-line pattern shown is that obtained when there is no friction between the surface of the metal and the anvils.

across the face of the anvil but is a maximum at the centre, the mean value being higher than the value $P = 2k$ given above.

Deformation of an 'ideal' plastic metal by a flat punch

The only problems of plastic indentation which have been rigorously solved are those involving two-dimensional deformation. We shall therefore first consider the deformation of a metal by a hard flat punch which is infinitely long and of uniform width d (Fig. 15). The metal is assumed to be ideally plastic, of yield stress Y. The punch is assumed to be undeformable and the friction between the face of the punch and the metal is considered to be negligibly small. When the load is first applied to the punch the stresses set up in the metal will be given by the elastic equations. The shear stresses at the edge of the punch will be very high. If, in fact, the edges of the punch are perfectly sharp, the shear stresses reach infinite values as soon as the smallest loads are applied. Consequently, even for the smallest applied loads the metal at the regions A and B will be in a state of incipient plasticity (Fig. 15). At other parts of the metal, however, the conditions are not sufficient to produce plastic flow, so that the overall yielding of the metal will be small and will be determined essentially by its elastic properties. As the load on the punch is increased, the region over which the metal becomes plastic increases in size, until finally the whole of the material around the punch is in a state of plasticity and the indenter begins to penetrate the metal on a relatively large scale. The determination of the pressure on

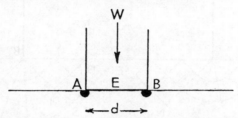

Fig. 15. Two-dimensional deformation (plane strain) of an ideal plastic metal by a flat punch of width d. The onset of deformation occurs at the edges A and B.

the face of the punch when this occurs was first derived by Prandtl (1920), and we may indicate how this may be obtained by the method of slip-lines.

It may be shown that the slip-line pattern which satisfies the plastic equations and the boundary conditions for stress and displacement is of the form drawn in Fig. 16. The four isosceles triangles, CMA, ANE, ESB, and BTG, represent domains in which the slip-lines are straight lines making an angle of 45° with the free face of the metal and the punch. The triangles on either side of the punch are connected to those beneath the punch by curved slip-lines which are arcs of circles with A and B as centres. The slip-lines orthogonal to these are the straight lines radiating from A and B. The region in which the material is flowing plastically is defined by the boundary $CMDNESFTGC$. We may now follow a typical α slip-line starting at the point G_1. As we saw above, since the normal principal stress Q at this point is zero, $p = k$. Consequently the equation along this slip-line may be written

$$p + 2k\phi = k,$$

where ϕ is the deviation of the slip-line from the direction $G_1 T_1$. Along $G_1 T_1$ there is no change in the value of p since $G_1 T_1$ is a straight line. At T_1, however, the slip-line begins to curve until the point S_1 is reached. The slip-line has now turned through an angle $\phi = -\frac{1}{2}\pi$, so that for the point S_1 we may write

$$p - 2k\tfrac{1}{2}\pi = k.$$

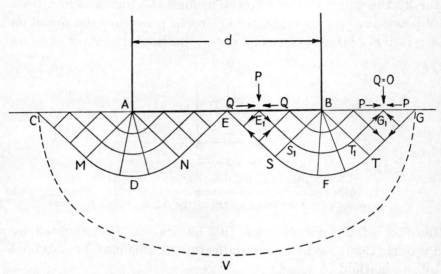

FIG. 16. The slip-line pattern in an ideally plastic metal deformed by a flat punch when large-scale deformation occurs. Plane strain (Hill, 1950).

No further change occurs in the value of ϕ as the slip-line is traversed from the point S_1 to the point E_1 so that at E_1

$$p = k + 2k\tfrac{1}{2}\pi.$$

We now determine the principal stresses at E_1. The stress normal to the surface, which is now P, is given by

$$P = p + k = 2k(1 + \tfrac{1}{2}\pi),$$

whilst the transverse stress, which is now Q, is given by

$$Q = p - k = 2k\tfrac{1}{2}\pi.$$

This means that the metal under the punch at the point E_1 is subjected to a normal pressure equal to $2k(1+\frac{1}{2}\pi)$ and to a transverse compressive stress of amount $2k\frac{1}{2}\pi$. Similar treatment shows that the same result is obtained for all points along the surface of the punch. Consequently the normal pressure P is uniform over the face of the punch and is given by

$$P = 2k(1 + \tfrac{1}{2}\pi). \tag{7}$$

This analysis was first given by Prandtl (1920) and Hencky (1923). For annealed mild steel where Tresca's criterion holds, $2k = Y$. For most other metals where the Huber–Mises criterion holds, $2k = 1\cdot15Y$. Thus in general, full-scale plastic flow and consequent indentation occur when the pressure P_m across the face of the punch is given by

$$P_m = 2\cdot6Y \text{ to } 3Y. \tag{8}$$

This relation has been approximately verified by Nadai (1923) and by other workers. It may be noted that two-dimensional deformation of the type discussed above is similar to that occurring in the rolling of wide sheets of metal, and Orowan (1944) has shown that the pressures involved are, in fact, given approximately by equation (8).

The region in which the material flows plastically is shown by the boundary $CDEFGC$ (Fig. 16). Within this region the elastic strains may be neglected compared with the large plastic strains which occur. Some plastic flow may also occur in the material adjacent to the boundary $CDEFG$, but in this region the plastic strains are of the same order of magnitude as the elastic strains. Outside the region CVG the metal is deformed elastically and the strains produced are always small (see Hill, 1950).

The effect of friction

In the above derivation it is assumed that there is no friction between the face of the indenter and the metal, so that the metal is perfectly free to slip laterally at the interface in the course of the indentation process. Under these conditions the pressure is uniform across the face of the punch and has a value of about $3Y$. If there is appreciable friction between the surfaces it will increase the pressure necessary to produce indentation. This follows because in the course of the indentation process there is slip of metal along the face of the indenter. Since the metal is more confined in the centre (region E, Fig. 16) than at the edges of the indenter (regions A and B, Fig. 16), the resultant pressure will be higher in the centre than at the edge. The flow pattern will also be modified and the overall mean yield pressure will be

higher. The detailed treatment to allow for friction is complicated, but some idea of the order of the effect may be obtained
by considering a much simpler case, described by Siebel (1923).

Let us consider the plastic yielding of a two-dimensional strip
of metal between hard flat anvils (Fig. 17). We again assume
that there is no flow in a direction normal to the plane of the

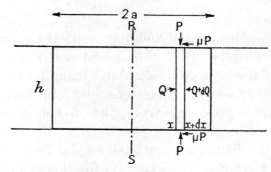

FIG. 17. Effect of friction on the deformation of a two-
dimensional strip of metal between flat anvils.

paper (plane strain). The strip has a width $d = 2a$ and a thickness h, and it has a constant elastic limit or yield stress Y. If
the surfaces between the anvils and the strip were perfectly
frictionless the pressure P_0 necessary to produce plastic yielding
would be uniform across the face of the anvils and would have
the value
$$P_0 = Y \quad \text{or} \quad P_0 = 1{\cdot}15Y, \tag{9}$$

depending on whether the Tresca or the Huber–Mises criterion
is adopted.

On account of the finite friction between the surfaces the
pressure will be increased and we may calculate the effect in the
following way. Because of symmetry the metal will flow sym-
metrically on either side of the central plane RS. Consider any
element, distant x from the centre, thickness dx. This element
is forced to the right and is subjected to a frictional force at the
interfaces which acts to the left. If P is the normal pressure on
this element there is a frictional force at each anvil-element
interface of $\mu P \, dx$ per unit length, where μ is the coefficient of
friction, so that the total force is $2\mu P \, dx$. This is just equal to

the force produced by the horizontal component of pressure Q within the metal. The resultant force is $h\,dQ$.

Hence $\qquad\qquad\qquad h\,dQ = -2\mu P\,dx.$ $\qquad\qquad$ (10)

The element is thus subjected to a transverse pressure Q and a vertical pressure P. This produces a shear stress in the

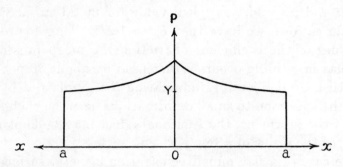

FIG. 18. Pressure distribution over the face of the anvils in Fig. 17 for an ideal plastic metal for which the yield stress is Y.

element equal to $\frac{1}{2}(P-Q)$, and the condition for plastic flow according to both the Tresca and the Huber–Mises criterion is simply that
$$P-Q = \text{constant}.$$

Hence $\qquad\qquad dP-dQ = 0 \quad\text{or}\quad dP = dQ.$

Equation (10) becomes $h\,dP = -2\mu P\,dx.$

Integration yields
$$P = A e^{(2\mu/h)(a-x)}. \qquad\qquad (11)$$

We may reasonably assume that at the edge of the anvil where $x = a$ the friction has no effect, so that the pressure is the same as that occurring in the absence of friction (P_0), so that $A = P_0$. Alternatively we may say that at the edge of the anvil Q must be zero, since there is no normal stress applied to the free surface. Consequently at this boundary, P must be equal to Y (or $1\cdot15Y$) and hence equal to P_0 (see equation (9)). Hence $A = P_0$.

Hence $\qquad\qquad\qquad P = P_0\,e^{(2\mu/h)(a-x)}.$ $\qquad\qquad$ (12)

The pressure distribution follows the curve shown in Fig. 18. The maximum, which occurs at the centre, has a value of

$P_0\,e^{(2\mu a/h)}$. If the exponential power is not too large this may be expanded to give $P_0\{1+(2\mu a/h)\}$. Consequently the mean pressure across the face of the anvil becomes approximately

$$P_m = P_0\left(1+\frac{\mu a}{h}\right). \tag{13}$$

If $h = a$ and $\mu = 0\cdot2$ (a typical value for metal surfaces with some lubrication), we have that $P_m = 1\cdot2P_0$. Thus for reasonable values of the coefficient of friction the mean pressure at which plastic yielding occurs is increased by about 20 per cent. In practice, by deliberately lubricating the anvils and by subjecting the specimen to small deformations between each application of the lubricant, the frictional effect may be kept much smaller, of the order of a few per cent. or less.†

In the case of a flat punch penetrating a metal surface, the flow pattern is appreciably different from that applying above. Nevertheless, the effect due to friction will probably be of the same order of magnitude. That is to say, with surfaces as normally used in practical tests, the pressure at which the punch begins to penetrate the metal surface will not be more than a few per cent. greater than the theoretical value of about $3Y$.

Deformation by a flat circular punch

Although the solution for the problem of a two-dimensional punch is exact, it is not possible to solve rigorously the problem of a flat circular punch. However, Hencky (1923) and more recently Ishlinsky (1944) have shown that the pressure at which a circular punch will penetrate the surface of a metal is of the same order as that given in equation (8).

In order to solve this problem, which by rigorous treatment is intractable, both Hencky and Ishlinsky have used the Haar–Karman criterion of plasticity described above. The slip-line pattern obtained by Ishlinsky is shown in Fig. 19. It is seen

† This derivation is only valid if $\mu P < k$. If the friction is high enough to make $\mu P = k$, slip will occur within the skin of the metal near the anvil, rather than at the interface itself. The pressure to produce plastic flow will be higher, and for $a = h$, P_m will have a value greater than $1\cdot5P_0$.

that the slip-lines leave the free surface of the metal at an angle
of 45° and end up at the surface of the punch at an angle of 45°.
The slip-lines have therefore all turned through an angle of 90°,
although the path through which they have travelled is different
from that given in Fig. 16. We might therefore expect that
the normal pressure across the face of the punch would be uni-

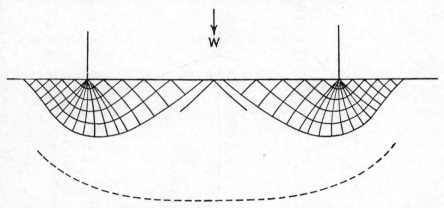

Fig. 19. The slip-line pattern for a flat circular punch penetrating an ideally
plastic material. The solution derived by Ishlinsky is based on the Haar–
Karman criterion of plasticity. The broken line is an approximate representation of the elastic-plastic boundary, corresponding to the line CVG in Fig. 16.

form, as for the two-dimensional model, and have a value of
$P = 2k(1+\tfrac{1}{2}\pi)$. Because Ishlinsky has used the Haar–Karman
criterion of plasticity, however, and has applied it to a three-
dimensional problem, the hydrostatic pressure p along the slip-
lines no longer follows the simple relations given in equation (6).
The value of p depends on the path travelled by the slip-line.
As a result the normal pressure over the face of the punch is not
uniform even for a frictionless punch, but is somewhat higher
at the centre than at the edges. This is shown in Fig. 20. The
mean pressure for plastic penetration P_m is, however, not widely
different from the value $3Y$, where Y is the yield stress of the
metal. (The detailed calculation by Ishlinsky gives a value
$P_m = 2.84Y$.) This relation has also been approximately
verified.

If, of course, there is friction between the punch and the metal,
the yield pressure, that is the mean pressure to produce plastic

yielding, will be higher. In most indentation tests carried out under practical conditions the surfaces are covered with thin films of grease so that the coefficient of friction will generally have a value of about $\mu = 0.2$. Consequently the increase in the yield pressure due to friction will be small. The same probably applies to spherical indenters and to shallow pyramidal

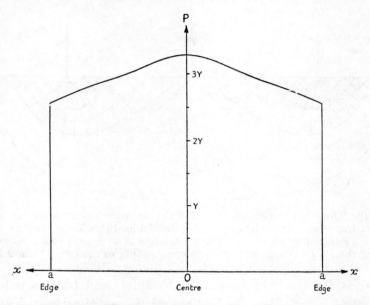

FIG. 20. Pressure distribution over the face of a flat circular punch penetrating an ideally plastic metal of yield stress Y (see Fig. 19).

or conical indenters if some lubrication is present. If, however, the surfaces are thoroughly freed of lubricant films the friction may reach values of the order of $\mu = 1$, and in such cases the effect of friction on the yield pressure may be very much more marked. Indeed, if the friction is too high, slip may occur within the metal itself rather than at the interface between the indenter and the metal. On the other hand, if the indenter is made of diamond the frictional effects will be less marked and more reproducible, since for unlubricated surfaces the coefficient of friction of diamond sliding on most metals (unlubricated) is of the order of $\mu = 0.1$ to 0.15 and this value is not greatly affected by the presence of lubricant films.

REFERENCES

BRAGG, L. (1949), *Proc. Camb. Phil. Soc.* **45**, 125. Also *Nature* (1942), **149**, 511.

HAAR, A., and VON KARMAN, T. (1909), *Nachr. d. Gesellschaft d. Wissensch. zu Göttingen*, 204.

HENCKY, H. (1923), *Z. ang. Math. Mech.* **3**, 250.

HILL, R. (1950), *The Mathematical Theory of Plasticity*, Oxford. This is an extremely valuable contribution to the theory of plastic deformation. It also discusses many practical metal-deforming processes.

——, LEE, E. H., and TUPPER, S. J. (1947), *Proc. Roy. Soc.* A, **188**, 273.

HUBER, A. T. (1904), *Czasopismo techniczne, Lemberg.*

ISHLINSKY, A. J. (1944), *J. Appl. Math. Mech.* (*U.S.S.R.*) **8**, 233. An English translation by D. Tabor has been published by Ministry of Supply, A.R.D. (1947), Theoretical Research Translation No. 2/47.

LUDWIK, P., and SCHEU, R. (1925), *Stahl u. Eisen*, **45**, 373; see also LUDWIK, P. (1909), *Elemente der tech. Mech.*, Berlin.

VON MISES, R. (1913), *Nachr. d. Gesellschaft d. Wissensch. zu Göttingen, Math.-phys. Klasse*, 582.

MOHR, O. (1900), *Zeits. d. Vereines Deutsch. Ingenieure*, **44**, 1; see also (1914), *Abhandlungen aus dem Gebiete der technischen Mechanik*, Ernst & Sohn, Berlin.

NADAI, A. (1931), *Plasticity*, McGraw-Hill, New York.

OROWAN, E. (1943), *Proc. Instn. Mech. Engrs.* **150** (4), 140.

PRANDTL, L. (1920), *Nachr. d. Gesellschaft d. Wissensch. zu Göttingen, Math.-phys. Klasse*, 74.

SIEBEL, E. (1923), *Stahl u. Eisen*, **43**, 1295.

TAYLOR, G. I. (1934), *Proc. Roy. Soc.* A. **145**, 362.

TRESCA, H. (1864), *C.R. Acad. Sci., Paris*, **59** (2), 754.

CHAPTER IV

THE DEFORMATION OF METALS BY SPHERICAL INDENTERS: IDEAL PLASTIC METALS

Initial Plastic deformation

WE may now consider the deformation of an 'ideal' plastic metal, of yield stress Y, by a hard spherical indenter of radius r (Fig. 21 a). The friction between the indenter and the metal

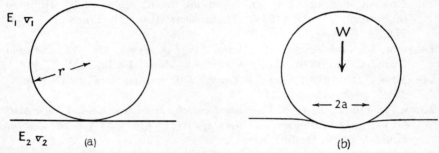

FIG. 21. Deformation of a flat surface by a hard sphere. So long as the deformation is elastic the radius of the circle of contact is proportional to $W^{\frac{1}{3}}$.

surface is again assumed to be negligibly small. When the load is applied to the indenter the metal surface and the indenter will both deform elastically according to the classical equations of Hertz (1881). The region of contact is a circle of radius a (see Fig. 21 b) as given by the equation

$$a = \left\{ \tfrac{3}{4} Wgr\left(\frac{1-\sigma_1^2}{E_1}+\frac{1-\sigma_2^2}{E_2}\right)\right\}^{\frac{1}{3}}, \tag{1}$$

where W is the load applied, E_1 and E_2 are Young's moduli of the indenter and the metal respectively, and σ_1 and σ_2 are the corresponding values of Poisson's ratio. Since Poisson's ratio has a value of about 0·3 for most metals, this gives

$$a = 1{\cdot}1\left\{\frac{Wgr}{2}\left(\frac{1}{E_1}+\frac{1}{E_2}\right)\right\}^{\frac{1}{3}}. \tag{2}$$

Thus the projected area A of the indentation is proportional to $W^{\frac{2}{3}}$ and the mean pressure P_m over the region of contact is

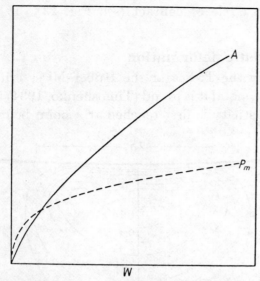

FIG. 22. Elastic deformation of a flat surface by a sphere: the area of contact A is proportional to $W^{\frac{2}{3}}$ and the mean pressure P_m is proportional to $W^{\frac{1}{3}}$.

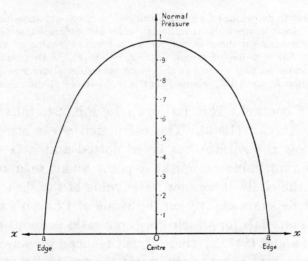

FIG. 23. Pressure distribution over circle of contact when a flat surface is deformed elastically by a sphere.

proportional to $W^{\frac{1}{3}}$ (Fig. 22). We may note that the pressure (or normal stress) across the circle of contact is not uniform, but at any point distant x from the centre of the indentation has the value $P = P_0(1-x^2/a^2)^{\frac{1}{2}}$, where P_0 is the pressure at the centre of the circle of contact (see Fig. 23). It follows that $P_0 = \frac{3}{2}P_m$.

Onset of plastic deformation

If we apply the Tresca or the Huber–Mises criterion to the stresses in the metal it is found (Timoshenko, 1934) that the condition for plasticity is first reached at a point below the actual

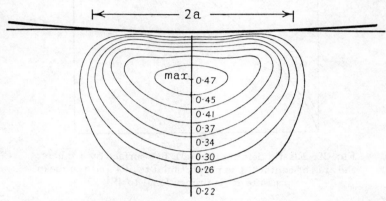

FIG. 24. Elastic deformation of a flat surface by a sphere, showing the maximum shear stress in the bulk of material below the deformed surface (Davies, 1949). The maximum shear stress occurs below the centre of the circle of contact and has a value of about $0 \cdot 47 P_m$, where P_m is the mean pressure. Plastic deformation first occurs at this point when the shear stress $= \frac{1}{2}Y$, i.e. when $P_m \approx 1 \cdot 1 Y$, where Y is the yield stress of the metal.

surface of contact. This is shown in Fig. 24, taken from a paper by Davies (1949). The calculated shear stress in the metal below the surface has been plotted and it is seen that the maximum value occurs at a point about $0 \cdot 5a$ below the centre of the circle of contact. The value of the shear stress at this point depends slightly on the value of Poisson's ratio, but for most materials for which Poisson's ratio is about $0 \cdot 3$ it has a value of about $0 \cdot 47 P_m$, where P_m is the mean pressure over the circle of contact. Since at this point the two radial stresses are equal, the Tresca and the Huber–Mises criterion both indicate that plastic flow will occur when the shear stress equals $\frac{1}{2}Y$, i.e. when $0 \cdot 47 P_m = 0 \cdot 5Y$. This means that plastic deformation commences at this region when

$$P_m \approx 1 \cdot 1 Y. \tag{3}$$

Provided the mean pressure is less than this, the deformation remains completely elastic, and on removing the load, the surface and the indenter return to their original shape. As soon as P_m reaches the value of $1 \cdot 1Y$, however, some plastic deformation

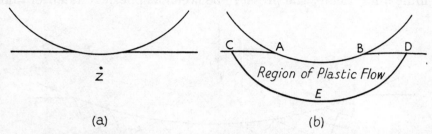

(a)   (b)

FIG. 25. Plastic deformation of an ideally plastic metal by a spherical indenter: (a) the onset occurs at a localized region Z when $P_m \approx 1 \cdot 1Y$, (b) at a later stage the whole of the material around the indenter flows plastically.

occurs at the region Z (Fig. 25 a). The rest of the metal is deformed elastically, so that on removing the load the amount of residual deformation is very small.

Complete or full plastic deformation

As the load on the indenter is increased, the amount of plastic deformation around the indentation increases and the mean pressure steadily rises until the whole of the material around the indentation is in a state of plasticity (Fig. 25 b).† It is not easy to define the stage at which this occurs, and the simplest approach is to say that it is reached when the yield pressure varies little with further increase in indentation size. This difficulty of defining the 'fully' plastic stage is also inherent in the theoretical treatment; here again it is assumed that the stage of 'full' plasticity has been reached, meaning that the plastic slip-line field covers the whole of the region around the indenter. Even so, the theoretical analysis even for the fully plastic stage cannot be carried out rigorously since the axially symmetrical

† Suitable annealing and etching techniques usually reveal a deformation pattern which differs in some details from that shown in Fig. 25 b. In particular, deformation does not appear to extend far beyond the edge of the indentation AB. It is possible that the plastic strains in the regions BD and AC may be too small to be revealed by metallographic techniques. A more detailed study of this deformation process would be very valuable.

problem in plasticity presents certain difficulties which are insuperable. Recently, however, Ishlinsky (1944) has extended the earlier work of Hencky (1923) and has found that by using the Harr–Karman criterion of plasticity it is possible to determine analytically the pressure between a spherical indenter and

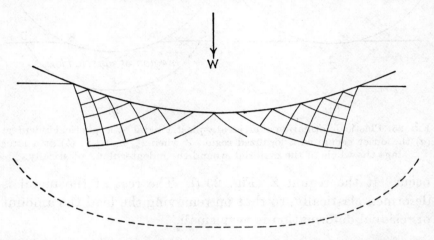

FIG. 26. Part of the slip-line pattern obtained by Ishlinsky for a spherical indenter deforming an ideally plastic metal. The analysis which is based on the Haar–Karman criterion of plasticity does not deal with the displacement of the deformed material. The broken line is an approximate representation of the elastic-plastic boundary, corresponding to the line *CED* Fig. 25 *b* and the line *CVG* in Fig. 16.

the indentation under conditions of 'full' plasticity. It must be borne in mind that this analysis is based on a physical assumption, the Haar–Karman criterion, which is not strictly valid, but the type of error involved does not appear to be serious and the result may be considered to be a good approximation. Part of the slip-line pattern obtained by Ishlinsky is shown in Fig. 26 and the pressure distribution in Fig. 27. It is seen that, as with the flat circular punch, the pressure over the surface of the indenter is not uniform over the region of contact but is somewhat higher in the centre than at the edges. However, the mean pressure P_m, that is the load divided by the projected area of the indentation, has a value of about $2 \cdot 66Y$. Ishlinsky also finds that P_m is not markedly dependent on the size of the indentation. It varies somewhat with the depth of penetration, and the analysis indicates that it is greater for a *flat* circular punch (the limiting case for small penetration) than for a spherical indenter submerged to an appreciable depth. As we shall see below, experiments suggest that P_m should *increase* somewhat with the depth of penetration, rather than decrease. However, the effect is not

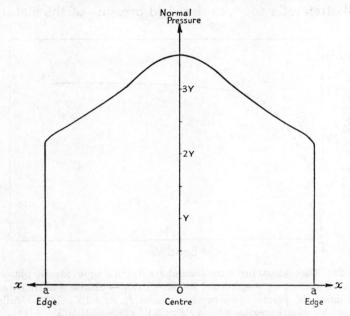

Fig. 27. Pressure distribution over the indentation formed by a spherical indenter in an ideally plastic material of constant yield stress Y (see Fig. 26).

very marked and Ishlinsky's calculations for the flat punch give a value $P_m = 2\cdot84Y$, which is only a few per cent. different from that calculated for the spherical indenter. Indeed it would seem from Ishlinsky's analysis that over a wide range of experimental conditions $P_m = 2\cdot6$ to $2\cdot9Y$. If, of course, there is appreciable friction between the indenter and the metal it will lead to some increase in the value of P_m.

Pressure–load characteristic

We may thus expect that the pressure–load characteristic of a spherical indenter penetrating an 'ideal' plastic body will follow the curve shown in Fig. 28. The portion OA represents the initial elastic deformation where the mean pressure is proportional to $W^{\frac{1}{3}}$. The point L corresponds to the onset of plastic deformation which occurs when $P_m = 1{\cdot}1Y$. The dotted portion LM represents a transitional region as the amount of plastic flow increases, whilst MN represents the condition of 'full' plasticity where P_m is of the order of $3Y$. For convenience we shall often refer to P_m as the yield pressure of the metal, and

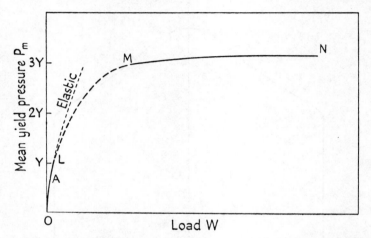

Fig. 28. Theoretical pressure–load characteristic of an ideally plastic metal deformed by a spherical indenter: OA = elastic deformation, L = onset of plastic deformation where $P_m \approx 1{\cdot}1Y$, MN = fully plastic region where $P_m \approx 3Y$. (Tabor, 1950.)

we may note that according to Fig. 28 it increases from a value of about $1{\cdot}1Y$ to a value of about $3Y$ as the deformation passes from the onset of plastic deformation to a 'fully' plastic state.

We may at once test the conclusion that the yield pressure under fully plastic conditions is approximately equal to $3Y$. This is most simply carried out by making large Brinell indentations in metal specimens that have been highly worked, so that they are incapable of appreciable further work-hardening and may be considered to possess a constant yield stress Y. Some typical results are given in Table III, (Tabor, 1948), the yield stress Y being found from 'frictionless' compression experiments. In these experiments it is found that the value of P_m increases slightly with the depth of the indentation, presumably on account of the increased confinement of the displaced material (see also Bishop, Hill, and Mott, 1945). This observation appears to be at some variance with the theoretical conclusions of Ishlinsky discussed above, but in any case the effect is small and it is clear that, to a first approximation, 'full' plastic yielding occurs when

$$P_m = cY, \tag{4}$$

where c is nearly a constant and has a value of about 3.

TABLE III

Metal (Work-hardened)	Y (kg./mm.²)	P_m (kg./mm.²)	$c = P_m/Y$
Tellurium–lead alloy .	2·1	6·1	2·9
Aluminium . .	12·3	34·5	2·8
Copper . . .	31	88	2·8
Mild steel . . .	65	190	2·8

These results show at once that for fully work-hardened materials the yield pressure is essentially independent of the load and of the size of the indentation. This is, of course, the same as saying that the Meyer index n for work-hardened materials is 2. In what follows we shall discuss a number of further conclusions that have a direct bearing on the practical measurement of hardness.

Range of validity of Meyer's law

It is instructive to carry out a series of indentation measurements on a highly worked metal starting from the smallest loads at which indentations are visible. The result of a plot of P_m against W is shown in Fig. 29, for highly worked mild steel, the dotted line referring to the calculated elastic deformation. It is seen that the curve has all the characteristics of the theoretical curve given in Fig. 28; the yield pressure begins to deviate from the elastic curve at a value of $P_m \approx 1 \cdot 1 Y$ and gradually rises until it reaches the value of about $2 \cdot 8Y$. This effect is shown more strikingly in Fig. 30, where the diameter d of the indentation has been plotted against the load W on logarithmic ordinates. The portion OL corresponding to the elastic region is a straight line of slope 3. (This is calculated from the elastic equations.) The portion LM is gently curved, but may be considered to be composed of approximately straight portions of slopes 2·7 and 2·25. Finally the portion MN is a straight line of slope 2. It is at once evident that MN is the range over which Meyer's law is valid for the highly worked steel. Here the Meyer

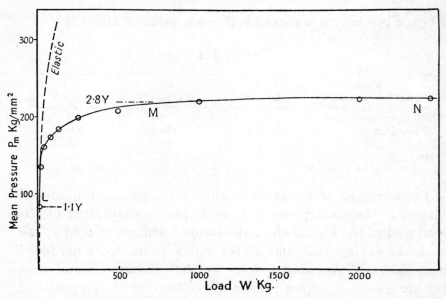

FIG. 29. Experimental pressure–load characteristic of indentations formed in work-hardened mild steel by a hard spherical indenter. Yield stress of steel $Y = 77$ kg./mm.² Ball diameter = 10 mm. The broken line is the theoretical result for elastic deformation. (Compare Fig. 28.)

index n is constant and has the value 2. As the load is decreased the value of the Meyer index gradually increases, until when the deformation becomes completely elastic it reaches the upper value of 3.

It is not difficult to estimate the load above which Meyer's law is valid. Suppose W_L is the load at which the onset of plastic deformation occurs (point L, Fig. 29) and W_m is the load at which full plasticity is first attained (point M, Fig. 29). If the curve between L and M followed the elastic relation, the ratio of W_m to W_L would be approximately $(2·8Y/1·1Y)^3$, i.e. about 20. On account of the more gradual slope of the curve during the growth of the plastic region the ratio is between 100 and 200, say 150. We may readily calculate the value of W_L from the elastic equations, since from equation (2) we may express P in terms of W and then put $P = 1{\cdot}1Y$. The relation is

$$W_L = 13{\cdot}1P^3r^2\left(\frac{1}{E_1}+\frac{1}{E_2}\right)^2,\qquad\qquad(5)$$

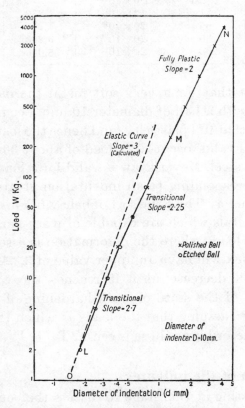

Fig. 30. Indentation of work-hardened mild steel (see Fig. 29) plotted on a log W–log d curve. Yield stress of steel $Y = 77$ kg./mm.² Ball diameter $= 10$ mm. Portion OL is the calculated elastic curve (slope 3), L is the onset of plastic deformation, LM is a transition range and MN the range over which full plastic deformation occurs (slope 2).

where r is the radius of the indenter. Putting $r = 0{\cdot}5$ cm. and $E_1 = 20.10^{11}$ dynes/cm. for a steel ball of diameter 10 mm. (as used in the Brinell test), we obtain the results given in Table IV for highly work-hardened metals.

TABLE IV

Metal (Work-hardened)	Y (kg./mm.²)	E_2 (dynes/cm.²)	W_L to give $P_m = 1 \cdot 1 Y$ (g.)	Approximate load for Meyer's law to be valid (kg.)
Tellurium–lead	2·1	$1 \cdot 6 \times 10^{11}$	2	0·3
Copper	31	12×10^{11}	230	35
Mild steel	65	20×10^{11}	1,200	180
Alloy steel	130	20×10^{11}	9,800	1,500
Very hard steel	200	20×10^{11}	35,000	5,200

It is apparent that for a very soft metal the onset of plastic deformation with a ball of diameter 10 mm. occurs at a load of about 2 g., so that full plasticity (and hence the load above which Meyer's law is valid) occurs at a load of about 300 g. Similarly for very hard steel, Meyer's law is valid for a load above about 5,200 kg., corresponding to an indentation greater than about 3 mm. in diameter. The same sort of behaviour may be expected to hold for metals which are capable of work-hardening. Thus at very small loads where the deformation is essentially elastic the Meyer index will have an upper value of 3. At higher loads the index will decrease until it reaches the constant value characteristic of the state of work-hardening of the metal. It is reasonable to assume that the loads at which this occurs will be of the same order as those given in Table IV.

Deformation of the indenter

It is interesting at this stage to discuss the conditions under which the indenter itself may be permanently deformed in the course of the indentation process. It is clear that for soft metals the indenter will be deformed only elastically, but for harder metals some permanent deformation may occur.

Suppose the metal has a yield pressure at full plasticity of B corresponding to a yield stress Y, where $B \approx 2 \cdot 8Y$, and suppose the indenter has a yield pressure B_i corresponding to a yield stress Y_i when again $B_i \approx 2 \cdot 8Y_i$. To a first approximation the yield pressure, or Meyer hardness, is the same as the Brinell hardness, so that we may call B and B_i the B.H.N.s of the metal and indenter respectively. As the load on the indenter is increased, plastic deformation of the metal will begin to occur at a mean pressure of about $1 \cdot 1Y$. If $Y_i > Y$, this pressure will be less than $1 \cdot 1Y_i$, so that the stress will be insufficient to produce any plastic deformation of the indenter. As the load is increased further, the mean pressure between the indenter and the metal increases until it reaches a value of about $2 \cdot 8Y$, but it does not appreciably exceed this value. If, therefore, the pressure is not to be sufficient to produce even the onset of plastic deformation of the indenter, it must be less than $1 \cdot 1Y_i$, that is $2 \cdot 8Y < 1 \cdot 1Y_i$ or $Y_i > 2 \cdot 5Y$ or $B_i > 2 \cdot 5B$. Thus the indenter should be at least two and a half times as hard as the metal under examination. This means that with the balls commonly used in Brinell hardness measurements possessing a hardness number or yield pressure of about 900 kg./sq. mm., metals of hardness number or yield pressure greater than about 400 kg./sq. mm. should not be used. This is in close agreement with the empirical convention generally adopted in Brinell hardness measurements.

These conclusions are of quite general validity. If local plastic deformation is produced in one metal by a second metal, the second metal should be at least two and a half times as hard as the first if it is not to suffer some permanent deformation itself.

Brinell hardness measurements of very hard metals

We may also consider here the problems involved in the hardness measurement of very hard materials. In making Brinell measurements of hard materials it is customary to use a 10-mm. ball and a load of 3,000 kg. The results obtained up to a hardness value of about 300 are satisfactory and are in agreement with those obtained by other methods, particularly those involving the use of pyramidal or conical indenters. For reasons which will be apparent in Chapter VII, the diamond pyramidal indenter as used in the Vickers hardness test gives hardness values which are reasonably reliable even for the hardest metals. With the Brinell test, however, the hardness values above about 300 are less than the Vickers values and the divergence becomes very much more marked for very hard materials. This has been attributed to the deformation of the indenter in the Brinell test and, for very hard metals, spherical indenters of sintered carbide or even of diamond may be used. Although this tends to give higher Brinell hardness values for the same load, there is still an appreciable difference between them and the Vickers hardness values. This is shown in Table V.

TABLE V

Hardness Values for Balls of Different Materials

Vickers pyramid hardness (kg./sq.mm.)	Brinell hardness numbers. 10-mm. ball; 3,000-kg. load		
	Steel ball	Tungsten carbide ball	Diamond ball 1-mm. ball; 30-kg. load
1,200	780	870	..
1,000	710	810	900
750	600	680	..
550	495	525	530
400	388	388	400
305	302	302	304
130	130	130	130

It is clear that with metals above 900 B.H.N., the steel ball will undergo more deformation than the metal itself and the hardness of the ball will, in fact, set an upper limit to the hardness value obtained. This does not, however, explain why there is still an appreciable difference when the hardness of the metal is only 550. Nor does it explain the differences observed with the balls of tungsten carbide (B.H.N. $\approx$ 1,500) or of diamond (B.H.N. $\approx$ 6,000†). In what follows we shall show that the low Brinell hardness values observed with hard metals may be explained on the assumption that, at the standard specified load of 3,000 kg., the indentation is not large enough to reach the stage of 'full' plasticity. Consequently the yield pressure does not reach the full value of about $2 \cdot 8Y$, but lies between $1 \cdot 1Y$ and $2 \cdot 8Y$. With the Vickers test, however, the deformation always involves the same degree of plasticity, so that the hardness value is more or less independent of the size of the indentation. There is not sufficient detailed experimental data fully to substantiate

† Deduced from experiments with the Knoop indenter. See p. 101.

this view, but it is interesting to follow this suggestion and estimate the magnitude of the effect to be expected.

We assume that the metals under consideration are fully work-hardened and that their $P_m - W$ characteristic is essentially the same as that observed with the work-hardened mild steel in Fig. 29. From Fig. 29 we may follow the growth of the plastic region and tabulate the increase of the yield pressure P_m in terms of the load. This is given in Table VI a, while in Table VI b the same results have been converted so that the load is expressed as the ratio W/W_L, where W_L is the load necessary to produce the onset of plastic deformation; the yield pressure is expressed as the ratio P_m/P_N, where P_N is the pressure at full plasticity and corresponds to the hardness value under 'proper' conditions of measurement. The ratio P_m/P_N is thus approximately equal to the ratio

$$\frac{\text{B.H.N. observed}}{\text{true B.H.N.}}.$$

TABLE VI

Indentation of Work-hardened Mild Steel by 10-mm. Ball

	(a)		(b)
Load W (kg.)	P_m (kg./sq.mm.)	Ratio $\dfrac{W}{W_L}$	Ratio $\dfrac{P_m}{P_N}$
(W_L) 2	84	1	1 : 2·55
5	105	2·5	1 : 2·05
10	120	5	1 : 1·8
20	142	10	1 : 1·5
40	160	20	1 : 1·35
80	180	40	1 : 1·2
125	186	62	1 : 1·17
250	200	125	1 : 1·08
500	210	250	1 : 1·03
700	216	350	1 : 1
2,000	220	1,000	1 : 1

Here again it is seen that full plasticity and, therefore, the stage at which reliable hardness measurements may be made, occurs when the load exceeds 100 to 200 times that at which the onset of plastic deformation occurs.

Let us assume that the Vickers hardness values would be the

same as the Brinell hardness numbers if full plasticity was reached. There are general reasons for believing that this is approximately true. Consider a steel specimen for which the true hardness value is 1,000 kg./sq. mm. This corresponds to a value of $P_N = 2 \cdot 8Y$, so that the onset of plasticity $(1 \cdot 1Y)$ corresponds to a yield pressure of about 390 kg./sq. mm. With a 10-mm. ball of tungsten carbide for which $E_1 \approx 6 . 10^{12}$ dynes/ sq. cm. the onset of plastic deformation according to equation (5) will occur at a load of 90 kg. Thus, when in the hardness test a load of 3,000 kg. is used, the load is 33 times greater than that at which the onset of plasticity occurs. From Table VIb we see that for $W/W_L = 33$ the ratio $P_m/P_N \approx 1 : 1 \cdot 25$. The observed yield pressure will therefore have a value of

$$1000/1 \cdot 25 = 800 \text{ kg./sq. mm.}$$

Table V shows the observed value to be about 810.

With the diamond indenter $E_1 \approx 10 . 10^{12}$ dynes/cm., so that here the onset of plastic deformation occurs at a load of about 72 kg. The load in the hardness test is 42 times as large so that $P_m/P_N \approx 1 : 1 \cdot 18$. The observed yield pressure should therefore be of the order of $1000/1 \cdot 18 = 850$. Table V gives a value of about 900. Similar calculations have been carried out for the tungsten carbide and diamond indenters on other metals and for the steel indenter on metals of hardness less than 750 B.H.N. The results are summarized in Table VII.

TABLE VII

Calculated and Observed Brinell Hardness Values

'True' hardness value (kg./sq.mm.)	Brinell Hardness Number. 10-mm. ball; 3,000-kg. load					
	Steel ball		Tungsten carbide ball		Diamond ball	
	Observed	Calculated	Observed	Calculated	Observed	Calculated
1,200	780	..	870	890	..	930
1,000	710	..	810	800	900	850
750	600	620	680	670	..	680
550	495	500	525	520	530	535
400	388	390	388	400	400	400
305	302	305	302	305	304	305
130	130	130	130	130	130	130

It is seen that the agreement between the observed and calculated values is good. We should not expect better agreement
for two main reasons. Firstly, the Vickers hardness may not be
identical with the 'true' Brinell hardness value, particularly if
the metal is not fully work-hardened. As we shall see in the next
chapter, the indentation produced by a spherical indenter work-
hardens the specimen by an amount depending on the size of the
indentation, and for small indentations (such as occur with hard
metals) the effective increase in hardness may be less than that
produced by the Vickers indenter.

Secondly, the growth of the plastic region may not follow
exactly the course shown in Fig. 29. For example, the general
theory of plastic deformation indicates that the fully plastic
stage should not be dependent on the elastic constants of the
metal and indenter. On the other hand, the onset of plasticity
depends critically on the elastic constants. It would, in fact,
appear that the onset of plasticity determines the initial growth
of the plastic region. Thus to a first approximation both the
onset of plasticity and the growth of the plastic region, which
we have calculated above, depend on the elastic constants of the
surfaces. This means that for materials of the same yield stress
the higher the elastic constants the sooner the plastic process is
initiated and the smaller the load necessary to reach 'full'
plasticity. Once this stage is reached, however, the value of the
yield pressure depends only on the plastic yield stress of the
material and not on the elastic constants.

In spite of these reservations, it is clear that the results in
Table VII confirm the view that with the standard 3,000-kg.
load, full plasticity is not reached for harder metals and for this
reason low hardness values are observed. Tungsten carbide and
diamond have relatively high Young's moduli, so that the onset
of plasticity occurs at smaller loads than with a steel indenter.
Consequently with these indenters the deformation at a load of
3,000 kg. is nearer the fully plastic stage than with a steel
indenter. Thus even at a hardness value of 550, where the steel
indenter is not markedly deformed, it will give lower hardness
values than the tungsten carbide and diamond indenters.

We may also consider the deformation of the steel indenter itself (B.H.N. $\approx$ 900) if it is pressed against an infinitely hard, unyielding metal surface. If $P_N = 900$, $E_1 = \infty$, the onset of plastic deformation occurs at a load of about 40 kg. If the growth of the plastic region is similar to that observed in Fig. 29, the full load of 3,000 kg. is 75-fold bigger than the load at which the onset of plasticity occurs, so that $P_m/P_N \approx 1 : 1 \cdot 1$. The yield pressure at this load will therefore be about 800 kg./sq. mm. This means that if the steel indenter is used on very hard metals, the upper limit to the hardness observed at a load of 3,000 kg. will be about 800 kg./sq. mm. and will be determined by the hardness of the ball itself.

It follows from this discussion that if reliable Brinell hardness values are to be obtained with very hard metals, tungsten carbide or diamond indenters must be employed and loads much greater than 3,000 kg. ought to be used with 10-mm. balls. This is often impractical, and for this reason metals possessing a hardness above about 800 should be tested with the Vickers indenter (see Chapter VII).

An associated problem is that of the Meyer index of very hard materials. Even if the metal is fully work-hardened with a true value of $n = 2$, appreciably larger values of n may be obtained if the loads are not sufficient to bring the indentation to the fully plastic stage (see Fig. 30). It is possible that this effect often occurs with very hard steels. For example O'Neill (1926) quotes a $0 \cdot 4$ per cent. carbon steel which in the annealed state had a Brinell hardness of 181 kg./sq. mm. and a value of the Meyer index $n = 2 \cdot 24$. When heavily quenched the hardness increased to a value of nearly 560 kg./sq. mm. The Meyer index increased to a value of $n = 2 \cdot 38$, indicating that the quenched specimen is capable of greater strain-hardening than the annealed specimen. This high value of n may, however, only be due to the increased hardness of the metal and the inadequacy of the applied loads. It would be interesting to know if, with much larger applied loads, the specimen would still give high values of n.

Effect of surface roughness

The above conclusions may also be applied to a consideration of the effect of surface roughness. We consider a hard steel surface pressed against a flat surface of a softer metal covered with asperities. For simplicity we assume that the asperities

(a)   (b)

FIG. 31. Deformation of asperities by a harder surface: (a) hemispherical asperity deformed by a flat surface, (b) flat surface deformed by a hemisphere. The deformation processes are similar in both cases.

have tips of spherical shape and that the steel indenter is perfectly smooth and of large radius of curvature compared with the asperities. Thus the deformation at each asperity may be considered as occurring between a hard flat surface and a spherical softer surface (Fig. 31 a). The behaviour is essentially the same as that occurring between a hard spherical indenter pressing on to the plane surface of a softer metal (Fig. 31 b). We may therefore use equation (5) to calculate the loads necessary to initiate plastic deformation in asperities of specified radii of curvature. Typical results for various materials are given in Table VIII and it is seen that for surfaces of small radii of curvature (r) the onset of plastic deformation occurs at extremely

TABLE VIII

Metal	Approximate Brinell hardness (kg./sq. mm.)	Yield stress, Y (kg./sq. mm.)	Load at which onset of plastic deformation occurs ($P_m = 1 \cdot 1 Y$) (g.)			
			$r = 10^{-4}$ cm.	$r = 10^{-2}$ cm.	$r = 0.5$ cm.	$r = 1$ cm.
Tellurium–lead	6	2·1	8×10^{-8}	8×10^{-4}	2	8
Soft copper	55	20	$2 \cdot 5 \times 10^{-6}$	0·025	62	250
Work-hardened copper	90	31	$9 \cdot 0 \times 10^{-6}$	0·09	230	910
Work-hardened mild steel	190	65	$4 \cdot 7 \times 10^{-5}$	0·47	1,200	4,700
Alloy steel	350	130	$3 \cdot 8 \times 10^{-4}$	3·8	9,500	38,000

small loads.† Thus for an asperity of radius of curvature 10^{-4} cm. on a hard steel surface, plastic deformation commences at a load of less than 10^{-3} g., so that the asperity is in a state of 'full' plasticity at loads less than 0·1 g. With an asperity of radius of curvature 10^{-3} cm., which is larger than the irregularities usually occurring on smooth surfaces, the corresponding figures are 0·04 g. and 6 g. It follows that when the indenter is pressed on to the surface the minute irregularities which are always present on the surface will readily deform beyond their elastic limit. The indenter will in fact be supported by asperities that have flowed plastically until their area is sufficient to support the applied load.

Although plastic flow of the asperities occurs so readily, this does not necessarily mean that the underlying metal is also deformed plastically. For example, if the steel indenter has a diameter of 10 mm. ($r = 0·5$ cm.), the load necessary to initiate plastic flow in the bulk of the alloy steel will be of the order of 10 kg. Thus for smaller loads, although the asperities will deform plastically, the underlying metal will still deform elastically. Indeed the outer boundary of the plastically deformed asperities will be determined by the elastic yielding of the underlying metal. At higher loads, of course, plastic flow of the underlying metal will also occur so that at a load of the order of 1,500 kg. there will be full plastic flow on both a macroscopic and a microscopic scale. On the macroscopic scale, full plastic flow occurs under a yield pressure of approximately $3Y_b$, where Y_b is the yield stress of the bulk of the metal. The yield pressure of the asperities will in general be higher, largely as a result of further work-hardening of the asperities which may occur even if the specimen has already been highly worked. This effect will also be assisted by friction between the indenter and the tips of the asperities. Thus in general the asperities, which directly bear the load, will have a yield pressure that may be considerably higher than that of the bulk metal. The macroscopic indentation, however, will correspond to the yield pressure of the bulk

† A similar analysis has been given by Tagg (1947) for the contact between a steel pivot and a sapphire jewel.

of the metal itself. Thus the yield pressure calculated from the macroscopic indentation will provide a reliable measure of the hardness of the bulk of the metal however much the asperities themselves are work-hardened. For this reason, hardness measurements do not depend appreciably on the surface finish of the metal specimen. Similarly surface irregularities on the surface of the indenter will not appreciably affect the macroscopic deformation. Thus by using etched balls it is possible to obtain reliable measurements of the bulk deformation of the metal. This is particularly useful, as O'Neill has pointed out, in making indentation measurements at very small loads where the bulk deformation may be predominantly elastic.

The effect of surface irregularities is shown very strikingly in some experiments recently described by Moore (1948). In these experiments it was not convenient to use a spherical indenter. Instead a smooth cylindrical indenter was used and it was pressed into the surface of a work-hardened copper surface in which a series of fine parallel grooves had been cut. The cylinder was arranged with its axis parallel to the grooves and indentations made at various loads. Profilometer records of the resulting indentations are shown in Fig. 32. Fig. 32 a is for a light load where the tips of the asperities have been deformed plastically, while no plastic deformation has occurred in the underlying material. In Fig. 32 b there is slight deformation of the bulk material, whilst in Fig. 32 c for a very heavy load the bulk deformation has been very severe. It is interesting to note, however, that even in this case the irregularities retain their identity and are clearly visible even at the bottom of the indentation.

It is clear from Fig. 32 c that marked plastic deformation has occurred over the two ranges described above. For the asperities it has occurred at their tips where the effective yield pressure is appreciably greater than $3Y_b$. For the bulk material below the asperities flow has also occurred at a yield pressure of about $3Y_b$. The area of the tips supporting the load is approximately one-half the area of the macroscopic indentation, so that the yield pressure of the asperities is approximately double that of the bulk metal. With annealed materials this divergence will be even more marked. It is evident, however, that the region of plastic deformation on the macroscopic scale will not depend essentially on the properties of the asperities, but on the yield pressure of the bulk metal itself.

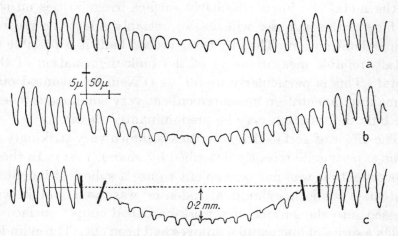

FIG. 32. Profilometer records of a grooved surface deformed by a hard cylinder placed with its axis parallel to the grooves: (a) light load, (b) heavier load, (c) very heavy load. For light loads the plastic deformation is restricted to the tips of the asperities. Only at heavier loads is the underlying metal deformed plastically, but even here the irregularities retain their identity.

'Piling-up' and 'sinking-in'

Consideration of the plastic region provides a simple explanation of the effects of piling up and sinking in. The region over which the major plastic deformation occurs is shown in Fig. 33. When the metal is displaced by penetration of the indenter it flows out between AC and BD, so that the material in this region is raised above the general level (Fig. 33 a). As the indenter descends there is also a marked lateral movement near A and B because of the increasing diameter of the indentation, so that the most marked piling up occurs around the edge of the indentation. This is the behaviour characteristic of an 'ideal' plastic material, i.e. of a highly worked metal.

If the material is in the annealed state, however, the behaviour is different. The early displacement of metal in the plastic region produces appreciable work-hardening and it becomes easier to displace the adjacent metal which lies deeper below the indentation. Consequently the displaced metal flows out at the region outside C and D (Fig. 33 b). Once this material has yielded it also work-hardens and further displacement of metal occurs at a still greater depth. As a result the material around

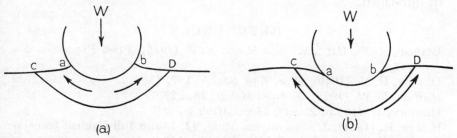

FIG. 33 (a) For highly worked metals the flow of metal around the indenter produces 'piling-up', (b) for annealed metals the displacement of metal occurs at regions at a small distance from the indenter so that 'sinking-in' occurs.

the indentation itself is left at a lower level than the material farther away from the indenter. This is the essential characteristic of the 'sinking in' observed with annealed metals.

'Strainless' indentation

The results discussed above have shown that when a metal is deformed by a spherical indenter the deformation is elastic until the mean pressure reaches a value of $1 \cdot 1Y$. This occurs whether the surface is initially flat or a portion of a sphere so that it also holds for a cavity or a preformed indentation. Thus for a highly worked metal the pressure which the metal can withstand without plastic deformation is simply $1 \cdot 1Y$. On the other hand, the yield pressure under the usual conditions of hardness measurements is of the order of $3Y$. Consequently the 'absolute hardness' as found by a strainless indentation method will be approximately one-third of the 'normal' hardness. This was observed by Mahin and Foss (1939) in their experiments on machined cavities.

For annealed metals we should expect the difference to be somewhat greater. The mean pressure during the strainless indentation will again be approximately $1 \cdot 1Y$, where Y is the yield stress of the annealed material. The 'normal' hardness of the metal will be greater than $3Y$ because of the appreciable work-hardening of the material around the indentation (see next chapter). Thus the ratio of the 'absolute' to the 'normal' hardness, in the method used by Harris, should be something less than one-third. Harris's values on the whole are of the order of one-third.

REFERENCES

BISHOP, R. F., HILL, R., and MOTT, N. F. (1945), *Proc. Phys. Soc.* **57**, 147.

DAVIES, R. M. (1949), *Proc. Roy. Soc.* A, **197**, 416.

HARRIS, F. W. (1922), *J. Inst. Metals*, **28**, 327.

HENCKY, H. (1923), *Z. ang. Math. Mech.* **3**, 250.

HERTZ, H. (1881), *J. reine angew. Math.* **92**, 156: a full English translation appears in *Miscellaneous Papers* (1896), London.

ISHLINSKY, A. J. (1944), *J. Appl. Math. Mech. (U.S.S.R)*, **8**, 233. An English translation has been published by Ministry of Supply, A.R.D. (1947), Theoretical Research Translation No. 2/47.

MAHIN, E. G., and FOSS, G. J. (1939), *Trans. A.S.M.* **27**, 337.

MOORE, A. J. W. (1948), *Proc. Roy. Soc.* A, **195**, 231.

O'NEILL, H. (1926), *Carnegie Scholarship Memoirs*, **15**, 233.

TABOR, D. (1948), *Proc. Roy. Soc.* A, **192**, 247.

—— (1950), *M.I.T. Summer Conference on Mechanical Wear*. Discussion, pp. 325–8.

TAGG, G. F. (1947), *J. Sci. Instr.* **24**, 244.

TIMOSHENKO, S. (1934), *Theory of Elasticity*, McGraw-Hill, New York.

CHAPTER V

DEFORMATION OF METALS BY SPHERICAL INDENTERS
METALS WHICH WORK-HARDEN

So far we have discussed the deformation of a metal by a spherical indenter when the metal has a yield stress Y that is essentially unaffected by the deformation produced by the indentation process itself. In practice this behaviour is characteristic of metals which are very highly worked, so that its practical application is somewhat limited. In this chapter we shall consider the extension of this analysis to metals which are annealed or only partially worked, so that they experience work-hardening during the indentation process. It may be said at the outset that a theoretical treatment of this problem has not been solved even for the case of two-dimensional deformation. We must therefore approach the problem in an empirical way.

Yield pressure as a function of the size of the indentation

When an indentation is formed by a spherical indenter, the material around the indentation is displaced and, in general, the yield stress Y will be increased. However, as we shall see below, the elastic limit will not be constant at every point around the indentation since the amount of deformation or strain will in general vary from point to point. We may, however, expect that when full plasticity is reached there will exist an average or 'representative' value of the elastic limit, say Y_r, which is related to the yield pressure P_m by a relation of the type $P_m = cY_r$, where c is a constant having a value of about 3. Making this assumption we may consider the way in which Y_r depends on the size of the indentation and hence derive a relation between P_m and the size of the impression.

Suppose the indentation has a chordal diameter d and a radius of curvature r_2. Since it is a portion of a sphere its shape is completely defined by the dimensionless ratio d/r_2. Then for all

indentations for which d/r_2 is the same, the amount of deformation or strain at the 'representative' region will be the same (if
the grain size of the material is sufficiently small as to be irrele-
vant). This follows because strain itself is a dimensionless
parameter which depends only on the *fractional* change of
dimensions and not on the absolute change. For example, if a
uniform cylindrical bar 1 in. long is extended by $\frac{1}{10}$ in., the
strain produced *and the increase in yield stress produced by that
strain* will be the same as that occurring in a bar 3 in. long
extended by $\frac{3}{10}$ in. We may therefore say that the strain ϵ_1
produced at the 'representative' region will be simply a function
of d/r_2. If D is the diameter of the indenter, r_2 is usually very
nearly equal to $D/2$, so that ϵ_1 is approximately a function of
d/D. We may therefore write

$$\epsilon_1 = f(d/D). \tag{1}$$

This equation simply means that geometrically similar indentations produce similar strain distributions. In particular the
strain produced at the 'representative' region, and hence the
'representative' yield stress Y_r, depend only on d/D. Consequently the mean pressure P_m which is equal to cY_r will depend
only on d/D. Thus in the most general terms we may write

$$P_m = \frac{4}{\pi} \frac{W}{d^2} = \psi\left(\frac{d}{D}\right), \tag{2}$$

where $\psi(d/D)$ is some function of d/D that still has to be determined (see Chap. II, equation $(6\,c)$). This is, of course, the
same as saying that geometrically similar indentations have the
same hardness whatever the absolute size of the indentation (see
Chap. II).

We may express this conclusion more formally. If the metal
is fully annealed, ϵ_1 is the total strain produced at the 'representative' region by the indentation process. If, however, the
metal has previously been cold-worked, we may consider it as
annealed material that has undergone an initial strain ϵ_0. As we
shall see later, we may to a first approximation add this strain
to that produced by the indentation. Hence the total strain

produced at the 'representative' region will be given by

$$\epsilon = \epsilon_0 + f(d/D). \tag{3}$$

We assume that the yield stress Y_r is a single-valued function of the strain, i.e. if the metal is subjected to a specified amount of strain the yield stress can have only one value determined by the stress–strain characteristics of the metal. We may write

$$Y = \phi(\epsilon), \tag{4}$$

so that the 'representative' value of the yield stress becomes

$$Y_r = \phi\{\epsilon_0 + f(d/D)\}. \tag{4a}$$

The yield pressure when full plasticity is reached is then given by

$$P_m = c\phi\{(\epsilon_0 + f(d/D)\}, \tag{5}$$

where c is a constant having a value of about 3. For a given metal this is, of course, identical with equation (2).

Co-ordination of results

One conclusion that immediately follows from equation (5) is that we may at once co-ordinate hardness measurements made with various loads and ball diameters on a given specimen. For a fixed metal, ϵ_0 is constant, so that if we plot P_m against d/D we should obtain a single monotonic curve for all the loads and all the ball diameters used. Some results by Krupkowski (1931) for annealed copper are plotted in Fig. 34. It is seen that all the points lie about a smooth curve for ball diameters ranging from 1 to 30 mm. For the very small indentations it is possible that the conditions are not those of full plasticity, but in any case the principle of geometrical similarity still holds. We may also note that the curve in Fig. 34 is of the same type as the stress–strain curve for annealed copper.

The yield pressure as a function of the stress–strain characteristic

We may now consider a more quantitative connexion between the mean pressure P_m and the yield stress of the material around the indentation. A convenient method of measuring the yield stress of a material is to determine its hardness using a pyramidal indenter possessing a large apex angle, as in the Vickers test. In this case, as we shall see in Chapter VII, the mean pressure on the indenter is almost independent of the size of the indentation, i.e. the hardness is almost independent of the load. If, therefore, we measure the Vickers hardness of a metal that has

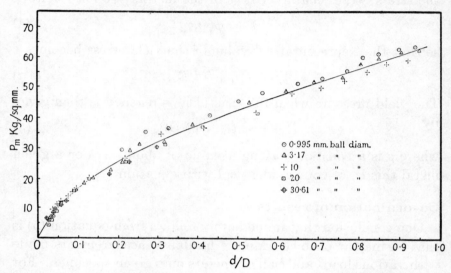

FIG. 34. Indentation of annealed copper for various loads and balls of various diameters. The plot of the mean pressure against the parameter d/D gives points which all lie on a single curve. Data from Krupkowski (1931).

been compressed or elongated by various amounts, we obtain a direct relation between the hardness number, the amount of deformation or strain, and the yield stress at any stage. This relation may then be used to determine the yield stress of any specimen of the metal (Tabor, 1948).

Relations of this type were determined for specimens of mild steel and annealed copper. Blocks of these metals were carefully compressed by various amounts between well-lubricated anvils and the yield stress at each stage of compression determined. The Vickers hardness numbers were also determined at each stage. Typical results for the mild steel specimens are shown in Fig. 35; the deformation is expressed as the change in length divided by the *compressed* length and corresponds to the fractional increase in the area of cross-section of the specimen. From Fig. 35 we may determine the deformation and yield stress of a specimen of mild steel by simply measuring its Vickers hardness. For example, if the Vickers hardness is 194, the specimen has experienced a deformation or strain of 13 per

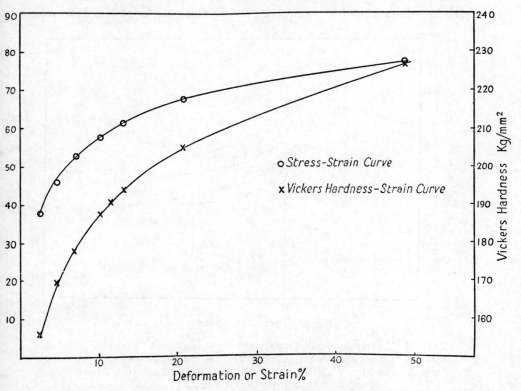

FIG. 35. Yield stress (circles o) and Vickers hardness (crosses ✕) of mild steel as a function of the deformation or strain. The specimen was deformed under compression and the strain ordinate is the areal strain.

cent. and its yield stress is 60 kg./mm.² By this means we may determine the yield stress of a metal when, for example, it has been deformed by a spherical indenter.

Brinell impressions of various sizes were made in the surface of mild steel and annealed copper specimens, and Vickers hardness measurements made at small loads (to give very small impressions) in the free surface of the specimen. From these measurements the yield stress of the deformed material in the free surface around the indentation and in the indentation was determined. Typical results for indentations of various sizes in mild steel are shown in Fig. 36. It is seen that the yield stress of the metal gradually rises as we approach the edge of the indentation. At the edge itself, the yield stress rises rapidly and then falls somewhat as we approach the centre of the indenta-

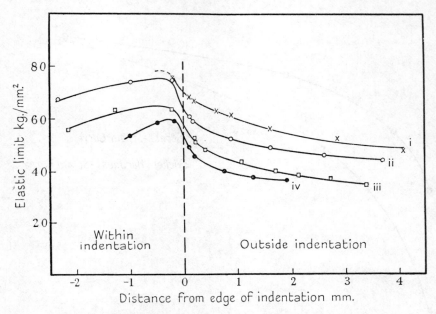

FIG. 36. Determination of the yield stress in the free surface around the indentation and in the indentation for impressions of various sizes formed in mild steel: (i) $d/D = 0·84$, (ii) $d/D = 0·69$, (iii) $d/D = 0·49$, and (iv) $d/D = 0·23$.

tion. There is also a variation in yield stress at various depths in the bulk of the material (O'Neill, 1934). It would therefore appear difficult to assign a 'representative' value to the yield stress of the whole material. Empirical tests suggest, however, that the yield stress at the edge of the indentation may be used as a 'representative' value for the whole of the deformed material around the impression. For example, we may compare the yield stress Y_e at the edge of the indentation with the mean pressure P_m involved in the formation of the indentation. Results for copper and steel are given in Table IX.

TABLE IX

Metal	Size of impression (d/D)	Y_e (kg./mm.²)	P_m (kg./mm.²)	Ratio P_m/Y_e	% deformation corresponding to Y_e
Annealed copper	0·27	10·5	27	2·6	5
	0·37	14	39	2·8	8
	0·5	16	44	2·8	9
Mild steel	0·23	51	132	2·6	6
	0·49	57	159	2·8	9
	0·69	63	161	2·6	15
	0·84	70	190	2·7	≈20

It is seen that over a wide range of indentation sizes, $P_m = cY_e$, where c has a value lying between 2·6 and 2·8. The last column of this table also shows that the deformation corresponding to Y_e is approximately proportional to the ratio d/D, i.e. if we express the deformation as a percentage

$$\epsilon_1 \approx 20d/D.$$

A simple example will make this relation clearer. Suppose we make an indentation of chordal diameter 5 mm. in a metal specimen with an indenter of diameter 10 mm. The ratio of d/D is equal to $\frac{1}{2}$. This means that the representative deformation is equivalent to a strain of 10 per cent. From the stress–strain curve of the metal we may determine the yield stress Y corresponding to a strain of 10 per cent. Then the mean pressure involved in producing the indentation will have a value of about $2 \cdot 8Y$.

This provides a direct means of comparing the stress–strain characteristics of a metal with the hardness curves. Fig. 37 shows the results obtained for annealed copper and mild steel. For the hardness curves the values of P_m have been plotted against the values of d/D. For the stress–strain curves the yield stress has been multiplied by a factor of 2·8 and plotted against the strain where the strain in per cent. has been made equal to $20d/D$. It is seen that there is close agreement between the hardness results and the stress–strain curves.

By analogy with the calculations described on pp. 51–4 we may expect that full plasticity will be reached for the mild steel when d/D is greater than about 0·1 and for copper at a smaller value. Consequently the whole of the range covered by the curves in Fig. 37 corresponds to a region of full plasticity in which c remains essentially constant.

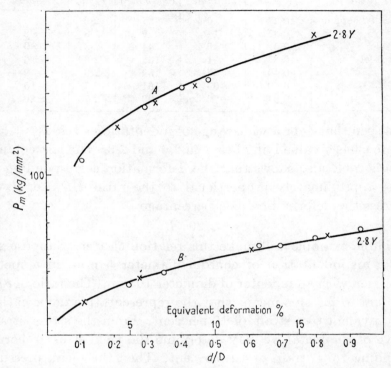

Fig. 37. Comparison of hardness measurements with the stress–strain curve. A, mild steel. B, annealed copper. o ✕ hardness values expressed as mean pressure over the indentation (Meyer hardness). —— stress–strain curve obtained from 'frictionless' compression experiments.

Yield pressure and stress–strain curves for deformed metals

We may extend this analysis to hardness measurements which have been carried out on specimens that have been deformed by various amounts (Tabor, 1948). Experiments show that the representative deformation is approximately additive to the initial deformation, i.e. at the edge of the indentation where the yield stress is Y_e the deformation may be written approximately as

$$\epsilon = \epsilon_0 + f\left(\frac{d}{D}\right).$$

In addition it is again found that $P_m = cY_e$, where c has essentially the same value as before. We should thus expect to obtain a series of P_m–d/D curves that have been displaced along the strain axis by amounts equal to the initial deformation of the specimen. An idealized curve is shown in Fig. 38 for a specimen that has been deformed by 0, 30, and 70 per cent. Experimental

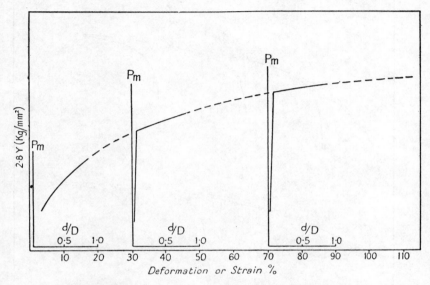

FIG. 38. Idealized curve showing the stress–strain curve (broken line) and the hardness values (continuous line) for specimens deformed by 0, 30, and 70 per cent.

results obtained for annealed copper and ordinary bright mild steel are shown in Fig. 39. The hardness values for copper are for annealed specimens that have been deformed in compression by 0, 9·6, 17·1, 29·6, and 41·5 per cent. The hardness values for mild steel are for specimens that have been deformed in compression by 0, 11·4, 22·1, and 35·7 per cent. It is again seen that there is reasonably close agreement between the hardness results and the stress–strain curves. It will be noted, however, that the hardness curves deviate from the stress–strain curves at those regions corresponding to smaller indentations, particularly for the steel specimen. This is largely due to the fact that for the work-hardened materials the condition of full plasticity is only reached for relatively large indentations, so that c can be considered constant only for indentations larger than a critical size. The discrepancy is also increased by the fact that the deformations are not exactly additive. Nevertheless, the general agreement is satisfactory.

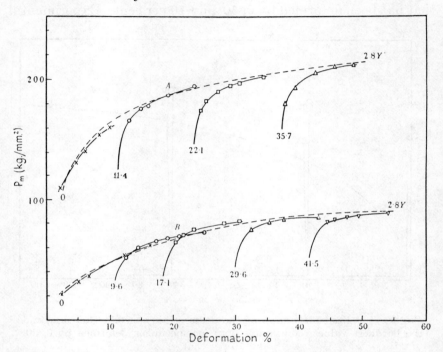

FIG. 39. Experimental results showing the stress–strain curve (broken line) and the hardness values (continuous line) for A, mild steel specimens; B, annealed copper specimens which have been deformed by various amounts.

Derivation of Meyer's laws

We may readily derive Meyer's laws for an annealed metal. Over an appreciable range of deformation, the yield stress of most metals may be expressed approximately as a simple power function of the deformation or strain ϵ, i.e.

$$Y = b\epsilon^x, \tag{6}$$

where b and x are constants (Nadai, 1931).

As we have seen, the deformation ϵ_1 at the edge of the indentation is, to a first approximation, directly proportional to the ratio d/D. We may, however, assume a more general relation and write

$$\epsilon_1 = \alpha(d/D)^y, \tag{7}$$

where α and y are constants. Hence for an annealed metal the representative value of the yield stress at the edge of the indentation will be given by

$$Y_e = b\epsilon^x = b\alpha^x(d/D)^{xy}. \tag{8}$$

Then, since $P_m = cY_e$, we have

$$P_m = \frac{4W}{\pi d^2} = cb\alpha^x\left(\frac{d}{D}\right)^{xy}, \tag{9}$$

so that

$$\frac{W}{d^2} = A\left(\frac{d}{D}\right)^z, \tag{10}$$

where $A = \frac{1}{4}\pi cb\alpha^x = $ constant and $z = xy = $ constant for the metal.

If for convenience we write $n = z+2$, this yields

$$W = \frac{Ad^n}{D^{n-2}}. \tag{11}$$

Thus for indentations made with balls of different diameters, $D_1, D_2, D_3...,$

$$W = k_1 d^n = k_2 d^n = k_3 d^n..., \tag{12}$$

where $k_1, k_2, k_3,...$ are given by

$$A = k_1 D_1^{n-2} = k_2 D_2^{n-2} = k_3 D_3^{n-2}.... \tag{13}$$

The two laws expressed by equations (12) and (13) (see Chap. II, equations (4) and (5)) were first deduced empirically by Meyer and are found to hold fairly accurately over a wide range of experimental conditions. Similar relations are also approximately valid for materials which have been cold-worked by various amounts. This follows because even for a material which has already been deformed by an amount ϵ_0, the stress–strain curve may still be represented, to a first approximation, by a relation of the type $Y = b_1\epsilon_1^{x_1}$, where b_1, x_1 are new constants and ϵ_1 is the additional strain. This applies, for example, to the partially annealed aluminium in Fig. 9. As we saw above, ϵ_1 is again roughly equal to $20d/D$, so that a relation identical to equation (11) again results. The value of n in the Meyer equations now corresponds to the new value x_1 of the stress–strain index.

The Meyer index and the stress–strain index

It is interesting to note that according to the experimental measurements described above, the power y in equation (7) is approximately unity, so that n in equations (12) and (13) is roughly equal to $(2+x)$. On the basis of earlier work by Kokado (1925), O'Neill (1944) has suggested that $n = 2+2x$. Most of the values given by O'Neill, however, are considerably nearer the relation $n = 2+x$, as the following table shows.

TABLE X

Comparison of Meyer Index n and Stress–Strain Index x. From Data by O'Neill (1944)

Metal	Typical values of Meyer index n (O'Neill)	$n-2$	Stress–strain index x	Kokado–O'Neill theory: $\dfrac{n-2}{2}$
Norris data:				
Mild steel A.	2·25	0·25	0·259	0·12
Yellow brass	2·44	0·44	0·404	0·22
Yellow cold drawn	2·10	0·10	0·194	0·05
Copper L	2·45	0·45	0·414	0·23
Stead data:				
Steel 1A	2·25	0·25	0·24	0·12
,, 2A	2·25	0·25	0·22	0·12
,, 4A	2·25	0·25	0·19	0·12
,, 6A	2·28	0·28	0·18	0·14
Schwarz data (Schwarz n-values):				
Copper				
(annealed)	2·40	0·40	0·38	0·20
(rolled)	2·12	0·12	0·04	0·06
Nickel				
(annealed)	2·50	0·50	0·43	0·25
(rolled)	2·14	0·14	0·07	0·07
Aluminium				
(annealed)	2·20	0·20	0·15	0·10

Brinell hardness and the ultimate tensile strength

We may now consider the relation between the Brinell hardness and the ultimate tensile strength of metals (Tabor, 1951). As we saw in Chapter III, the ultimate tensile strength T_m of an ideal plastic metal is essentially the same as the yield stress Y. Since P_m is approximately equal to $2 \cdot 8Y$ it follows that

$$P_m \approx 2 \cdot 8 T_m.$$

Hence the ratio $T_m/P_m \approx 1/2 \cdot 8 = 0 \cdot 36$. In Brinell hardness measurements, where the curved area instead of the projected area of the indentation is used to calculate the Brinell number B, the value of B is usually a few per cent. less than the corresponding value of P_m. Consequently the ratio of T_m/B will in general be a few per cent. higher than $0 \cdot 36$, say $0 \cdot 37$ to $0 \cdot 38$. We have, therefore, that for fully worked metals the ratio $T_m/B = 0 \cdot 37$ to $0 \cdot 38$ if T_m is expressed in kg./mm.² If T_m is expressed in tons/in.² this ratio becomes $0 \cdot 23$ to $0 \cdot 24$. This is close to the empirical values quoted in Chapter II (p. 17).

By a simple extension of our earlier conclusions we may derive a more general relation for the ratio T_m/B for metals of any degree of work-hardening. We consider the stress–strain curve of a tensile specimen where the true yield stress Y is plotted against the fractional increase in length ϵ. The type of stress–strain curve is similar to that used in the derivation of Meyer's law; we may therefore assume that

$$Y = b\epsilon^x, \tag{14}$$

where $x = n-2$. This curve is shown by the full curve in Fig. 40. From this curve we may readily calculate the apparent or nominal stress at any stage of the deformation process. If at any given point the fractional increase in length is ϵ, the length of the specimen is $1+\epsilon$. Since there is a negligible volume change during plastic deformation, the cross-section of the specimen will have been reduced by $1/(1+\epsilon)$, so that the nominal stress T will be given by $T = Y/(1+\epsilon)$. Thus the variation of T with ϵ becomes, from equation (14),

$$T = \frac{b}{1+\epsilon} \epsilon^x. \tag{15}$$

The value of ϵ at which this becomes a maximum is found by differentiating equation (15) with respect to ϵ and putting it equal to zero. Then

$$\frac{dT}{d\epsilon} = \frac{bx\epsilon^{x-1}}{1+\epsilon} - \frac{b\epsilon^x}{(1+\epsilon)^2} = 0,$$

or

$$x(1+\epsilon) = \epsilon.$$

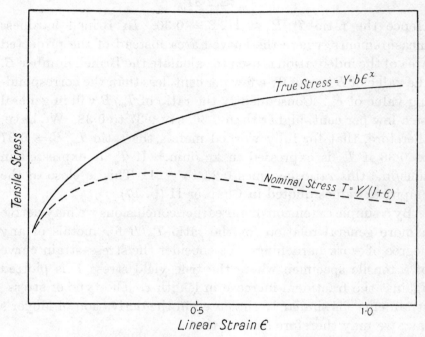

Fig. 40. Stress–strain curve under tension for a typical metal which work-hardens during deformation. Full curve, true stress; broken curve, nominal stress. Strain ordinate is linear strain.

Hence $$\epsilon = \frac{x}{1-x}. \tag{16}$$

It is of interest to consider two special cases. For fully work-hardened materials $x = 0$; for then $Y = b\epsilon^0 = b = $ constant. In this case $\epsilon = 0$. This means that the maximum tensile stress T_m occurs for zero strain, i.e. as soon as the metal yields plastically it begins to fail (see Chap. III, Fig. 11). This is, of course, the reason that for work-hardened materials $T_m = Y$. The second case of interest is when $x = 1$. Here $\epsilon = \infty$. This means that the metal work-hardens more rapidly than its cross-section decreases, so that its nominal tensile stress never reaches a maximum but steadily increases until the metal is reduced to zero cross-section. In practice, of course, this never occurs, for the index x never has a value greater than about 0·6.

Inserting the value of ϵ from equation (16) in equation (15) we obtain the maximum value of the nominal tensile stress T_m. We have

$$T_m = \frac{b}{1+x/(1-x)}\left(\frac{x}{1-x}\right)^x,$$

or
$$T_m = b(1-x)\left(\frac{x}{1-x}\right)^x. \tag{17}$$

We now determine the hardness value in terms of the stress–strain curve. If the indentation used in the hardness measurement has a size of $d/D = \frac{1}{2}$, the representative strain corresponds to $\epsilon = 10$ per cent. or $\epsilon = 0\cdot1$. Consequently the representative yield stress becomes
$$Y_e = b(0\cdot1)^x.$$

Thus the yield pressure P_m may be written

$$P_m = 2\cdot8b(0\cdot1)^x. \tag{18}$$

Since for an indentation of this size the ratio of the curved area of the indentation to the projected area is 1·07, the B.H.N. B will be about 7 per cent. less than P_m, i.e.

$$B = 2\cdot62b(0\cdot1)^x. \tag{19}$$

Combining equations (17) and (19) we find

$$\frac{T_m}{B} = \frac{1-x}{2\cdot62}\left(\frac{10x}{1-x}\right)^x. \tag{20}$$

The value of this ratio for values of x ranging from 0 to 0·6 is given in Table XI and plotted in Fig. 41.

TABLE XI

Ratio of T_m/B for an Indentation $d/D = \frac{1}{2}$

x	0	0·1	0·2	0·3	0·4	0·5	0·7
T_m/B	0·38	0·35	0·37	0·42	0·49	0·60	0·81

Since most Brinell hardness measurements are made with indentations ranging from about $d/D = 0.3$ to $d/D = 0.7$, similar calculations have been made for these values of d/D. It should be noted that for $d/D = 0.3$, $P_m = 1.024B$, whilst

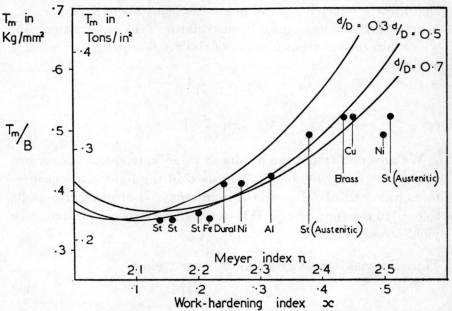

FIG. 41. Ratio of Brinell hardness B to ultimate tensile strength T_m as a function of the Meyer index n. The ratio depends slightly on the size of the indentation and the curves are drawn for three sizes, $d/D = 0.3$, 0.5, and 0.7.
The experimental results, taken from O'Neill, are shown as black circles.

$\epsilon = 0.06$. Similarly for $d/D = 0.7$, $P_m = 1.17B$, whilst $\epsilon = 0.14$. This leads to the following relations:

$$\text{For } \frac{d}{D} = 0.3, \qquad \frac{T_m}{B} = \frac{1-x}{2.73}\left(\frac{16.7x}{1-x}\right)^x. \qquad (20\,a)$$

$$\text{For } \frac{d}{D} = 0.7, \qquad \frac{T_m}{B} = \frac{1-x}{2.39}\left(\frac{7.14x}{1-x}\right)^x. \qquad (20\,b)$$

These results are also plotted on Fig. 41. It is seen that all the curves have the same characteristic. For values of the Meyer index ranging between 2 and 2·2 the ratio is roughly constant at a value of about 0·36 (in kg./mm.²). For higher values of the Meyer index the ratio increases fairly rapidly and reaches an upper value of about 0·5 (in kg./mm.²).

The results obtained from practical experiments are shown in circles for metals including drawn brass, annealed aluminium, and tool steel. These results are based on values given by O'Neill, but the data do not indicate the size of the indentations from which the practical ratio was deduced. Nevertheless, the general trend of results is similar to the theoretical curves and the points lie very close to the curve for $d/D = 0.7$. If we had taken the constant c in the relation $P_m = cY$ as being equal to 3 instead of 2·8, the observed results would lie around the theoretical curve for $d/D = 0.5$. We cannot expect better agreement than this, since the assumptions involved in this derivation are of an approximate nature. First, we have assumed that the stress–strain curve can be represented by a relation $Y = b\epsilon^x$. This holds only over a limited range. Secondly, we have assumed that $x = n-2$; thirdly, that the strain in the indentation used in the hardness measurements is $20d/D$. These assumptions are only approximate and small deviations lead to much larger errors in the ratio T_m/B. Nevertheless, as has been pointed out, the general trend of the experimental results and the actual values themselves are reasonably close to the theoretical curves.

It is of course clear that this treatment will not be valid if important structural changes occur in the tensile test which do not occur in the hardness measurements. This presumably applies to certain austenitic steels and also to materials containing serious flaws. For other metals, however, as the analysis shows, the hardness measurements may provide a simple and fairly reliable means of determining the tensile strength. (For a fuller discussion see Tabor, 1951.)

REFERENCES

KOKADO, S. (1925), *J. Soc. Mech. Engrs. Japan*, **28**, 257.

KRUPKOWSKI, A. (1931), *Rev. Métall.* **28**, 641.

NADAI, A. (1931), *Plasticity*, McGraw-Hill, New York.

O'NEILL, H. (1934), *The Hardness of Metals and Its Measurement*, Chapman and Hall, London.

—— (1944), *Proc. Instn. Mech. Engrs.* **151**, 115.

TABOR, D. (1948), *Proc. Roy. Soc.* A, **192**, 247.

—— (1951), *J. Inst. Metals*.

CHAPTER VI

DEFORMATION OF METALS BY SPHERICAL INDENTERS 'SHALLOWING' AND ELASTIC 'RECOVERY'

Cyclic deformation of 'recovered' indentations

WE have already noted that the permanent indentation left in a metal surface deformed by a hard spherical indenter has a larger radius of curvature than that of the indenting sphere. This effect has generally been ascribed to the release of elastic stresses in the specimen. It is at once evident that if this is indeed the case, it should be essentially reversible. That is to say, if the indenter is replaced in the 'recovered' indentation and the original load is applied, the surfaces should deform elastically, and on removing the load, the diameter and curvature of the 'recovered' indentation should remain unchanged.

We may now describe experiments which were carried out to test this (Tabor, 1948). A series of impressions were made with hard steel balls of various diameters on various metal surfaces, using loads ranging from 250 kg. to 3,000 kg. The diameters d of the impressions formed were measured after 1, 2, 3, and 5 cyclic applications of the load. The radius of curvature r_2 of the recovered indentation was also measured, using (a) a delicate profilometer, (b) a metallographic section across the diameter of the indentation. The values of d were reproducible to less than 1 per cent. The radii of curvature as determined by the profilometer method were reproducible to about 4 per cent. A few of these values were compared with those obtained from direct photomicrographs of the sections across the diameter of the indentation; the agreement was of the order of 1–2 per cent. For example, with a single application of load of 500 kg. on mild steel (10-mm. ball) the radius of curvature of the indentation by the profilometer method (mean of three determinations) was 0·595 cm. and by the direct contour method 0·605 cm. These values are typical. The results obtained in these experiments are given in Table XII.

TABLE XII

Metal	Radius of ball r_1 (cm.)	Load (kg.)	Dimensions of indentations (cm.) Number of applications of load			
			1	2	3	5
Brass . . .	0·952	500 d	0·27	0·270	0·270	0·270
		r_2	1·21	1·21	1·19	1·20
Aluminium alloy .	0·952	500 d	0·26	0·260	0·263	..
		r_2	1·15	1·16	1·16	..
Mild steel . .	0·5	500 d	0·183	0·185	0·186	0·192
		r_2	0·595	0·585	0·55	..
Hardened steel .	0·5	1,000 d	0·202	0·200	0·203	0·206
		r_2	0·677	0·677	0·68	0·652
	0·952	3,000 d	0·330	0·330	0·338	0·338
		r_2	1·39	1·37	1·31	1·37
	1·59	3,000 d	0·370	0·366	0·366	0·365
		r_2	2·80	2·77	2·84	2·71

It is seen that the indentation remains essentially unaltered in diameter and curvature after the second and third applications of the original load. This shows that the 'recovery' of the indentation when the load is removed is reversible and is therefore due to the release of elastic stresses.

'Shallowing' and elastic 'recovery'

Since the 'recovery' of the indentation is truly elastic, we may apply the classical laws of elasticity to the change in shape of the indentation. We idealize the condition of the surface of the metal after the indentation has been formed, and assume that it consists of a plane surface $XABY$ containing a depression of spherical form of radius of curvature r_2 and of diameter $d = 2a$ (Fig. 42 a).

When a hard steel sphere (radius of curvature r_1) is placed in the indentation, and a normal force of F dynes is applied, both surfaces are elastically deformed to a common radius of curvature r where $r_2 > r > r_1$ and the deformed surfaces finally touch over the boundaries of the indentation (Fig. 42 b). We assume that there is very little change in the diameter d during this deformation, an assumption which is generally accepted as being valid to within a few per cent. Then, according to Hertz's classical equations (Hertz, 1896) describing the elastic deformation of spherical surfaces, the relationship between d, r_1, and r_2 is given by

$$d = 2a = 2 \cdot 22 \left\{ \frac{F}{2} \frac{r_1 r_2}{r_2 - r_1} \left(\frac{1}{E_1} + \frac{1}{E_2} \right) \right\}^{\frac{1}{3}}, \tag{1}$$

where E_1, E_2 are Young's moduli for the indenter and the metal, and where we have assumed a value of $0 \cdot 3$ for Poisson's ratio.

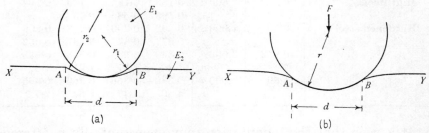

(a)   (b)

FIG. 42. Indentation produced by a spherical indenter: (*a*) 'recovered' indentation after removal of load, (*b*) after reapplication of load.

In a discussion of the derivation of this equation, Prescott (1927) has indicated that even if the surface $XABY$ is not a plane, the same equation will result. If, for example, the surface rises at the regions A and B as in Fig. 43 *a* or falls as in Fig. 43 *b*, the above equation is still valid, provided the projections or depressions at A and B are not too marked.

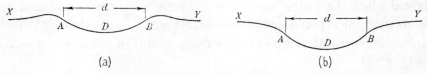

(a)   (b)

FIG. 43.

We apply this equation to the previous measurements of r_1, r_2, F, and d. We compare the value of d obtained from equation (1) with the observed value of the diameter of the recovered impression. The results are given in Table XIII. It is seen that the agreement between the last two columns is reasonably good, particularly as the accuracy in determining r_2 is not better than about 4 per cent.

TABLE XIII

Material	Assumed value of E_1 (dynes/cm.^{2})	Load (kg.)	Observed values (cm.)			Calculated d (cm.)
			r_1	r_2	d	
Brass	10×10^{11}	250	0·5	0·64	0·160	0·17
		500	0·952	1·21	0·27	0·26
Aluminium	7×10^{11}	250	0·5	0·66	0·178	0·18
alloy		500	0·952	1·15	0·26	0·30
Mild steel	20×10^{11}	500	0·5	0·605	0·183	0·20
Hardened	20×10^{11}	1,000	0·5	0·677	0·202	0·22
steel		3,000	0·952	1·39	0·330	0·36
		3,000	1·59	2·80	0·370	0·39

As a matter of interest, we make use of the observations of earlier workers.

1. *Profiles given by Batson (1918) for a ball of diameter 10 mm. and a load of 3,000 kg. on three types of steel.* We assume that $E_2 = 20 \times 10^{11}$ dynes/cm.^{2} for all the steels. The results are given in Table XIV.

TABLE XIV

Material	Observed values (cm.)			Calculated d (cm.)
	r_1	r_2	d	
Ni Cr steel . .	0·5	0·627	0·324	0·34
Manganese steel .	0·5	0·569	0·407	0·41
Rail steel . .	0·5	0·537	0·445	0·49

It is seen that the agreement between the observed and calculated values of d is reasonably good.

2. *Profiles given by Foss and Brumfield (1922) for a ball of 10 mm. diameter on various brasses.* The results are tabulated in Table XV.

TABLE XV

	Assumed value of E_2 (dynes/cm.^{2})	Load (kg.)	Observed values (cm.)			Calculated d (cm.)
			r_1	r_2	d	
Soft brass 1	9×10^{11}	3,000	0·5	0·518	0·555	0·7
Soft brass 2	9×10^{11}	500	0·5	0·521	0·330	0·38
Hard bronze 3	$7·5 \times 10^{11}$	3,000	0·5	0·527	0·497	0·66
Soft bronze 4	$7·5 \times 10^{11}$	500	0·5	0·557	0·276	0·28
Hard bronze 5	$7·5 \times 10^{11}$	3,000	0·5	0·531	0·499	0·63
Soft bronze 6	$7·5 \times 10^{11}$	500	0·5	0·566	0·302	0·28

It is seen that the agreement for the smaller loads (500 kg.) is good, whilst the agreement for the higher loads (3,000 kg.) is poor. This is probably because for the larger indentations the Hertzian equations are not accurately applicable. In addition, for soft metals the values of r_2 are very little different from r_1 for high loads, so that the errors introduced in calculating $(r_1 r_2)/(r_2 - r_1)$ may be very large. In the following table we take results where r is greater than 0·55 cm.

3. *Profiles given by Foss and Brumfield (1922) for a ball of 10 mm. diameter and a load of 3,000 kg. on various steels.* We assume that $E_2 = 20 \times 10^{11}$ dynes/cm.^{2} for all the steels. The results are given in Table XVI, and it is seen that the agreement between d (calculated) and d (observed) is close.

TABLE XVI

	Observed values (cm.)			Calculated d (cm.)
Metal	r_1	r_2	d	
0·5C–A	0·5	0·56	0·440	0·43
0·5C–W	0·5	1·03	0·26	0·25
0·9C–T	0·5	0·814	0·31	0·28
0·9C–W	0·5	1·372	0·240	0·23
MKD 455	0·5	0·568	0·349	0·36

These results show that in general the agreement between the observed and calculated values of d is reasonably good, particularly when r_2 is not too close to r_1, i.e. when the elastic 'recovery' is marked. It is, of course, true that as the calculation of d involves a cube root, the values of F, r_1, r_2, E_1, and E_2 are not very critical. Nevertheless, the agreement is consistent for a wide diversity of materials and experimental conditions.

Distribution of stresses

It is interesting to consider why the shape of the recovered indentation is essentially spherical. The Hertzian analysis shows that in the elastic deformation of solid bodies in contact there is only one pressure distribution which will deform a flat surface to a portion of a sphere or a spherical surface into a sphere of different radius. The normal stress distribution for this is shown in Chapter IV, Fig. 23, and is reproduced here in the full line in Fig. 44. Now the analysis by Ishlinsky shows that in the *plastic* deformation of metals by a spherical indenter the pressure distribution is given in Fig. 20, p. 42. This is reproduced in the broken line in Fig. 44. It is seen that the stress distribution is similar to the full line. Thus the stress distribution involved in

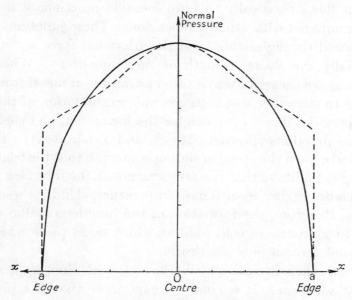

Fig. 44. Pressure distribution over circle of contact for a metal deformed by a spherical indenter. Full line, elastic deformation (Hertz); broken line, plastic deformation (Ishlinsky).

the plastic formation of the indentation is similar to that involved in the elastic deformation of spherical surfaces. Consequently when the load is removed the elastic stresses in the bulk of the metal are released and deform the indentation in a manner similar to that involved in the elastic deformation of spherical surfaces, where a spherical surface is deformed into a sphere of different curvature. It is for this reason that the recovered indentation is essentially spherical in shape. If the stress distribution during plastic deformation were widely different this would not be the case. This conclusion again suggests that though the premises on which the Ishlinsky analysis is based are not valid physically, the results obtained are close approximations to the true state of affairs.

Released elastic stresses and the adhesion of metals

These released elastic stresses play a very important part in the adhesion of metal surfaces. Investigations on the friction and surface damage of sliding metals show that when metal surfaces are placed together, the metal at the regions of real contact flows plastically to form metallic junctions which are large compared with atomic dimensions. These junctions must be sheared during sliding, and the frictional force is, in fact, essentially the shear strength of these junctions. When the junctions are sheared there is often a transfer of metal from one surface to the other and a microscopic examination of the surfaces provides direct evidence for the formation, and shearing, of these junctions (Bowden, Moore, and Tabor, 1941). This is observed even if the speed of sliding is so small that the frictional heating is slight, so that the temperature at the interface is not appreciably higher than room temperature. Under these conditions, therefore, the formation of the junctions is due essentially to a process of cold welding which takes place when the regions of contact flow plastically.

We may therefore ask why do the metals *not* show appreciable normal adhesion ? It is common experience that if a piece of copper is pressed on to a piece of steel, even if the surfaces are completely freed of lubricant films, they do not adhere together. Nevertheless, if they slide on one another there is ample evidence to show that intermetallic junctions are formed and sheared during the sliding process. Are these observations mutually contradictory ?

There are two answers to this question. First, the sliding process itself tends to break up surface contaminant films that may otherwise tend to interfere with the formation of metallic junctions. These films are not so readily penetrated under normal loading when sliding does not take place. Secondly, with harder metals the release of elastic stresses, as the load is removed, tends to break any junctions that may have been formed, so that by the time the load is completely removed there may be no junctions left. This view is confirmed by the fact that with softer metals where the released elastic stresses are much smaller, and where the junctions are appreciably more ductile, marked adhesion may be observed. If, for example, a clean steel ball is pressed on to the surface of a freshly cleaned bar of indium with a force of say 3 kg. it is found that the surfaces adhere very strongly and a normal force of about 3 kg. is required to separate the ball from the indium (see Fig. 45). Further, the sphere is

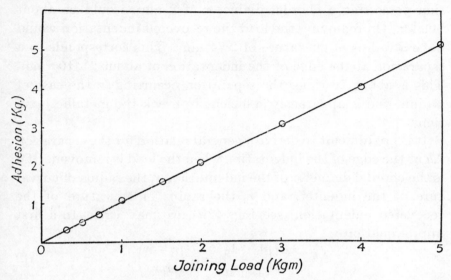

FIG. 45. Adhesion of a clean steel ball pressed into the surface of a freshly scraped specimen of indium. The force of adhesion is approximately equal to the original joining load. Diameter of steel ball ⅛″; time of loading 10 sec.

covered with small fragments of indium showing that the metallic junctions formed at the interface are at least as strong as the indium itself. For this reason the rupture occurs within the bulk of the indium specimen (McFarlane and Tabor, 1950).

It is interesting to estimate the change in shape of a typical indentation as a result of the released elastic stresses. If a steel ball of diameter 5 mm. is pressed on to the surface of a bar of indium with a force of 3 kg. it forms an indentation possessing a chordal diameter of about 2 mm. since the yield pressure of indium is only about 1 kg./mm.² If no adhesion occurred there would be relaxation of elastic stresses on removing the load and calculations show that the recovered indentation would have a radius of curvature of 2·51 mm. Thus the edge of the indentation would separate from the surface of the sphere by a distance of less than 10^{-4} cm. This is extremely small, and in practice this separation could readily be taken up by the ductility of the indium itself.

If a similar experiment were carried out on a steel bar of yield pressure 160 kg./mm.[2] the load necessary to produce an indentation possessing a chordal diameter of 2 mm. would be about 500 kg. On removing the load the recovered indentation would have a radius of curvature of 2·74 mm. This corresponds to a separation at the edge of the indentation of about 2×10^{-3} cm. This is about 20 times the separation occurring in the case of indium and is apparently sufficient to break the metallic junctions.

It is not difficult to derive a general relation for the separation Δh at the edge of the indentation when the load is removed. If d is the chordal diameter of the indentation, r_1 the radius of curvature of the indenter, and r_2 the radius of curvature of the recovered indentation (see Fig. 42 a), we may write, to a first approximation

$$\Delta h = \frac{d^2}{8}\left(\frac{1}{r_1} - \frac{1}{r_2}\right) = \frac{d^2}{8}\left(\frac{r_2 - r_1}{r_1 r_2}\right). \tag{2}$$

From equation (1) we may express $(r_2 - r_1)/(r_1 r_2)$ in terms of the Young's moduli of the indenter and the metal, E_1 and E_2 respectively, and in terms of the load W, where $Wg = F$. Remembering that the yield pressure P_m of the metal is given by

$$P_m = 4W/\pi d^2,$$

we obtain finally

$$\Delta h = 0.58 \times 10^8 P_m\, d\left(\frac{1}{E_1} + \frac{1}{E_2}\right), \tag{3}$$

where Δh and d are in cm., P_m is in kg./mm.[2], and E_1, E_2 are in dynes/cm.[2] It is evident from this equation that Δh is larger for harder metals, and as these are generally less ductile there is less chance of the metallic junctions remaining intact. Calculations such as these indicate the extent to which the elastic recovery of harder metals may nullify the adhesion which occurs when the load is originally applied. It is possible that in the sintering of metal powders, one of the reasons for carrying out the experiments at high temperatures is to release, by thermal means, the elastic stresses developed in the granules when the compact is formed. For a fuller discussion see McFarlane and Tabor (1950).

The processes involved in the Brinell test

We may now summarize the conclusions of the last three chapters and describe the processes involved in the Brinell hardness test. When a hard steel ball presses on to the surface of a metal specimen under examination the metal is first deformed elastically. As the load is increased, a stage is reached at which the maximum shear stress in the metal exceeds its elastic limit and the onset of plastic deformation occurs. This takes place when the mean pressure between the metal and the indenter reaches a value of about $1 \cdot 1Y$ (where Y is the yield stress or elastic limit of the metal), but the plastic deformation is confined to a small region below the centre of the region of contact. As the load is further increased, the mean pressure increases and the region of plasticity grows until the whole of the material around the region of contact is flowing plastically. At this stage the mean pressure reaches a value of about $3Y$. If the load is further increased the indenter sinks farther into the metal but the mean yield pressure remains approximately constant at a value of about $3Y$. This, of course, assumes that the material does not work-harden. If, as is generally the case, the metal work-hardens during the course of the indentation process, the effective value of Y may be considerably higher than the value of Y at the initial stage of the deformation. Consequently there are two simultaneous factors involved in the Brinell hardness test. The first is the transition from the onset of plastic deformation to 'full' plasticity as the mean pressure increases from $1 \cdot 1Y$ to $3Y$. The second is the increase in Y itself as the indentation process proceeds.

Most Brinell hardness measurements are carried out in the range where 'full' plasticity is reached, so that the main factor influencing the increase in yield pressure with load is the work-hardening characteristic of the metal. The extent to which the metal is work-hardened depends on the deformation produced by the indentation itself. This may be expressed in terms of the size of the indentation. Thus the effective deformation produced by an indentation of chordal diameter d, when the indenter has a diameter D, is approximately equal to $20d/D$, where the deformation is expressed as a percentage strain. In this way the increase in yield pressure with the size of the indentation may be directly correlated with the stress–strain characteristic of the metal. This analysis explains the well-established empirical laws of Meyer and also shows that the work-hardening index x is related to the Meyer index n by the relation $n \approx x+2$.

When equilibrium has been reached in the formation of an indentation, and plastic flow has come to an end, the whole of the load is borne by elastic stresses in the material. If the load is removed there is elastic 'recovery' of the indentation, with a corresponding change in its shape. If the indenter is re-applied to the recovered indentation with the same load the surfaces deform elastically until they just fit over the diameter of the original impression. The elastic stresses now reach the limits that the deformed material around the indentation can stand. If the load is removed or reduced, there is, as we have seen, a release of elastic stresses. If it is further increased, the stresses exceed the elastic limit and further flow of the metal occurs. There is a further increase in the size of the indentation and consequently in the amount of work-hardening and the process continues until the stresses, which are distributed over a larger indentation, again fall within the increased elastic limit.

REFERENCES

BATSON, R. G. (1918), *Proc. Instn. Mech. Engrs.* **2**, 463.

BOWDEN, F. P., MOORE, A. J. W., and TABOR, D. (1943), *J. App. Phys.* **14**, 80.

FOSS, F., and BRUMFIELD, R. (1922), *Proc. Amer. Soc. Test. Mat.* **22**, 312.

HERTZ, H. (1881), *J. reine angew. Math.* **92**, 156; see also *Miscellaneous Papers* (1896), London.

McFARLANE, J. S., and TABOR, D. (1950), *Proc. Roy. Soc.*, A **202**, 224.

PRESCOTT, J. (1927), *Applied Elasticity*, Longmans, London.

TABOR, D. (1948), *Proc. Roy. Soc.* A **192**, 247.

CHAPTER VII

HARDNESS MEASUREMENTS WITH CONICAL AND PYRAMIDAL INDENTERS

Conical indenters

A CONICAL diamond indenter for hardness measurements was first introduced by Ludwik in 1908. He used a cone having an included angle of 90° and defined the hardness as the mean pressure over the *surface* of the indentation. Thus if, for a

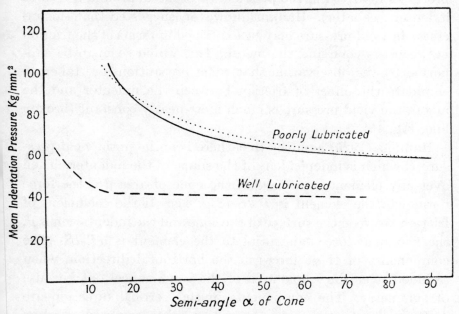

FIG. 46. Indentation of work-hardened copper by a conical indenter. Full line, poorly lubricated surfaces; broken line, well-lubricated surfaces (Bishop, Hill, and Mott, 1945). The dotted line is calculated from equation (2) assuming a value for the coefficient of friction of $\mu = 0.2$.

load W, the diameter of the impression is d, the surface area of the impression is $\pi d^2\sqrt{2}/4$, so that the Ludwik hardness number H_L is given by

$$H_L = \frac{4W}{\sqrt{2}\,\pi d^2}. \tag{1}$$

As we have seen in Chapter II, this pressure has no real physical significance. The true pressure P between the indenter and the indentation, if there is no friction between the surfaces, is again given by the ratio of the load to the *projected* area of the indentation, i.e. $P = 4W/\pi d^2$. It follows, therefore, that the Ludwik hardness number is simply $1/\sqrt{2}$ of the mean yield pressure P. Experiments show that the Ludwik hardness is practically independent of the load though it depends on the angle of the cone. This is shown in Fig. 46 for a hard steel cone indenting work-hardened copper (Bishop, Hill, and Mott, 1945). It is seen that the more pointed the cone (i.e. the smaller the semi-angle) the larger the yield pressure. This may be partly due to the detailed processes involved in the plastic flow of metal around a conical indenter (see later). Hankins, however, suggested that this increase in yield pressure may be explained in terms of the friction between the cone and the metal. This would seem to be supported by the observation that when precautions are taken to eliminate the effect of friction between the indenter and the metal, the yield pressure is much more nearly constant (broken line, Fig. 46).†

Hankins (1925) assumes that there is an intrinsic yield pressure P which is independent of the shape of the indenter. Then over any element of surface of the cone of area dS, the force normal to the element is $P\,dS$ (Fig. 47). If the coefficient of friction between the surface of the cone and the indentation is μ, the frictional force tangential to the element is $\mu P\,dS$. The components of these forces in the horizontal direction, when summed over the whole area of the cone, cancel out because of symmetry. The components in the vertical direction are $P\,dS\sin\alpha$ and $\mu P\,dS\cos\alpha$, and these, when summed over the area of the cone, must be equal to the normal force W.

Hence

$$W = \int dW = \int (P\,dS\sin\alpha + \mu P\,dS\cos\alpha)$$
$$= P(1 + \mu\cot\alpha)\int \sin\alpha\,dS.$$

† The experimental results shown in the broken line, Fig. 46, were carried out with a lubricated indenter which was withdrawn from the indentation and replaced with fresh lubricant and the original load re-applied. The process was continued until, for any fixed load, there was no further increase in the size of the indentation.

But $dS \sin \alpha$ is the projection of the area dS on the section AB, so that the integral of $dS \sin \alpha$ is simply equal to the area of the section AB, i.e. $\pi d^2/4$. Hence

$$W = P(1 + \mu \cot \alpha)\pi d^2/4$$

or
$$P = \frac{4W}{\pi d^2}\left(\frac{1}{1 + \mu \cot \alpha}\right)$$

$$= P_0\left(\frac{1}{1 + \mu \cot \alpha}\right), \tag{2}$$

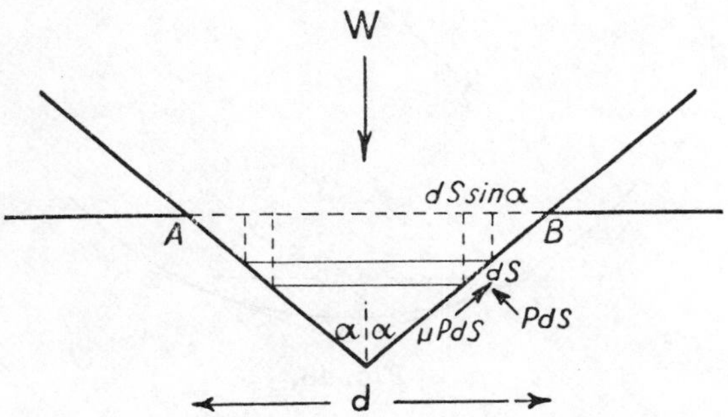

FIG. 47. Role of friction between a conical indenter and the deformed metal as treated by Hankins (1925).

where P_0 is the yield pressure when there is no friction. It is apparent that on this model the yield pressure is larger the smaller the value of α, i.e. the more pointed the cone. Assuming a value of $\mu = 0 \cdot 2$ and a value of $P_0 = 57$ kg./mm.^{2}, the results obtained using equation (2) are shown in the dotted line in Fig. 46. The agreement is good. It is doubtful, however, whether this model is satisfactory on general physical grounds. For example, when the cone becomes a flat cylindrical punch ($\alpha = 90°$), $P = 4W/\pi d^2$; this implies that the yield pressure is independent of the friction between the face of the indenter and the indentation. The full plastic treatment shows that this is not true. The analysis is too complicated to be given here, but the general trend has already been described in Chapter III.

Pyramidal indenters

The diamond pyramidal indenter was first introduced in hardness measurements by Smith and Sandland (1922) and was later developed by Messrs. Vickers–Armstrong, Ltd. It is also used in the Firth hardness apparatus. The indenter is in the form of a square pyramid and in the Vickers hardness machine the opposite faces make an angle of 136° with one another. The choice of this angle is based on an analogy with the Brinell test. For a Brinell test using a ball of diameter D it is customary to use indentations of diameters ranging between $0 \cdot 25D$ and $0 \cdot 5D$. The average of these is $0 \cdot 375D$. When tangents are drawn from

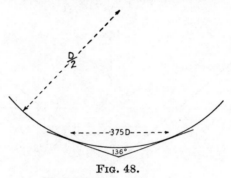

Fig. 48.

the points of contact of an impression of this diameter and the circumference of the indenter the included angle is 136° (see Fig. 48). The geometry of the indenter is such that the base of the pyramid has an area equal to $0 \cdot 927$ times the surface area of the faces. Since the Vickers hardness H_v is defined as the load divided by the surface area of the indentation, the yield pressure P is related to the Vickers hardness number by the relation

$$H_v = 0 \cdot 927P.$$

In making Vickers hardness measurements the lengths of the diagonals of the indentation are measured. If the mean value of this is d, and the indentation is square, the projected area of the indentation is $d^2/2$, so that the yield pressure is $2W/d^2$. Hence

$$H_v = 0 \cdot 927(2W/d^2).$$

The loads usually range from 1 to 120 kg. according to the hardness of the metal under examination, and the indentation usually has a diagonal length less than 1 mm. Smaller loads may also be used when a micro-hardness examination is needed, though this necessarily reduces the accuracy of measurement. The

Vickers hardness is expressed in kg./mm.^{2} and for normal use the accuracy may be better than $\frac{1}{2}$ per cent. Experiments show (see F. C. Lea, 1936) that the Vickers hardness number is independent of the size of the indentation and therefore of the load. In this it differs from the Brinell test, but for a given road

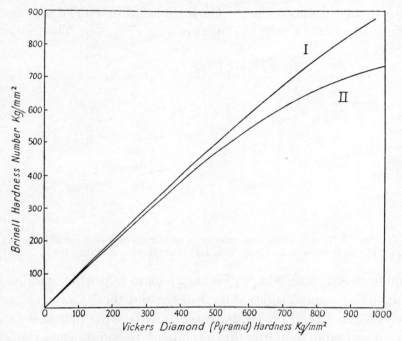

FIG. 49. Relation between Vickers and Brinell hardness values. Curve I, Brinell values obtained using a 10-mm. steel ball loaded to give an impression equal to 0·375 times the diameter of the ball. Curve II, Brinell values obtained using a 10-mm. steel ball under a constant load of 3,000 kg. Based on data given by Williams (1942).

the Brinell and Vickers numbers are generally very nearly equal, as Fig. 49 shows.

Although the indenter, being made of diamond, suffers very little deformation during the formation of the indentation, it is generally found that when the indenter is removed the impression is not a perfect square. For example, for annealed metals the impression has concave boundaries (pincushion appearance) corresponding to 'sinking-in' of the metal around the flat faces of the pyramid. For highly worked materials the indentation

has convex boundaries (barrel-shaped appearance) corresponding to 'piling-up' of the metal around the faces of the indenter.
Empirical corrections for these effects have been suggested.

The Knoop indenter

This indenter is a diamond pyramid in which the included
conical angles, subtended by the longer and shorter edges respectively, are 172° 30′ and 130° respectively (Fig. 50). The indenta-

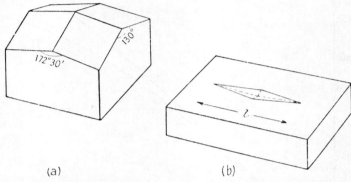

(a)   (b)

FIG. 50. (a) The diamond indenter used in Knoop hardness measurements,
(b) the indentation formed. This has a length seven times its breadth.

tion formed has the shape of a parallelogram in which the longer
diagonal is about 7 times as large as the shorter diagonal (Knoop,
Peters, and Emerson, 1939). Experiments show that there may
be considerable reduction in the shorter diagonal, when the load
W is removed, due to elastic 'recovery'. The longer diagonal l,
however, changes very little in length and is used as the basis
for the hardness measurement. From the geometry of the indenter and the length l, the projected area A of the 'unrecovered'
indentation is calculated and the Knoop hardness H_k is defined
as $H_k = W/A$. On the other hand, the change in dimensions of
the shorter diagonal may be used as a measure of the elastic
properties of the material.

In general the loads used in the Knoop tests vary from about
0·2 to 4 kg. with indentations of length about 0·1 mm. The
hardness values are not markedly dependent on the load and
are almost identical with the Vickers hardness numbers. However, the Vickers indenter penetrates the surface about twice as

far as the Knoop indenter for the same load so that, when operating at equal loads, the Vickers test is less sensitive to surface variations than the Knoop test. Indeed the shallowness of the Knoop indentation provides a means of examining the hardness of the uppermost surface layers of a metal. In addition the Knoop indenter often produces satisfactory indentations in materials such as glass which are not easily indentable by the Vickers pyramid or by a spherical indenter. It has even been possible to produce Knoop indentations in diamond, and the resulting hardness is about 6,000 kg./mm.² (Lysaght, 1946).

The indentation of an ideal plastic metal by a wedge-shaped indenter

The indentation of an ideal plastic solid by a conical or pyramidal indenter involves theoretical problems which have

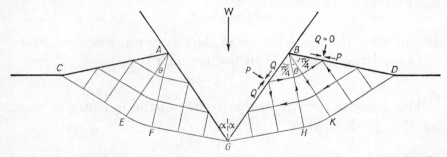

FIG. 51. Slip-line pattern for a two-dimensional wedge penetrating an ideally plastic material of yield stress Y (Hill, Lee, and Tupper, 1947). The pressure across the face of the indenter is uniform and has the value $P = 2k(1+\theta)$, where θ is the angle HBK in radians. This analysis allows for the displacement of the deformed material.

not yet been satisfactorily resolved. A rigorous solution for the two-dimensional model has, however, been obtained by Hill, Lee, and Tupper (1947). The two-dimensional indenter, of course, becomes a wedge, and the shape of the indentation is geometrically similar whatever its size. Consequently whatever the size of the indentation the flow pattern is unchanged and the pressure between the indenter and the indentation remains constant. The region over which large-scale plastic flow occurs is shown in Fig. 51, and experiments show that there is very good agreement between the theoretical flow pattern and that observed in practice. It should be noted that this analysis takes into account the displacement of the deformed material. It may be shown from geometric considerations that the angle θ in Fig. 51 is related to the semi-angle α of the wedge by the relation:

$$\cos(2\alpha-\theta) = \frac{\cos\theta}{1+\sin\theta}.$$

The variation of θ with α is shown in curve I, Fig. 52.

We may follow the progress of an α slip-line from the free surface to the surface of the indenter (Fig. 51). At the free surface $Q = 0$ so that $p = k$. Along the α slip-line we have (see Chap. III, equation (6))

$$p+2k\phi = \text{constant},$$

so that, starting from the free surface with $\phi = 0$, the constant must equal p. Hence

$$p+2k\phi = k.$$

In following the α slip-line from the free surface to the surface of the indenter, the angle turned through is $\phi = -\theta$. Hence

$$p = k+2k\theta.$$

The pressure normal to the surface of the indenter is given by $P = p+k$, so that

$$P = 2k(1+\theta).$$

The same value is obtained at all points along the surface of the indenter, so that if the friction at the interface is negligible the pressure P across the face of the indenter is uniform. The variation of P with θ and hence with α is shown in curve II, Fig. 52.

If we adopt the Huber–Mises criterion, for which $2k = 1{\cdot}15Y$, we have

$$P = 1{\cdot}15Y(1+\theta).$$

Thus when the semi-angle α of the wedge is 90° it becomes a two-dimensional flat punch for which

$$P = 2k(1+\tfrac{1}{2}\pi) = 2k(2{\cdot}57) = 2{\cdot}96Y.$$

As the angle α is reduced P steadily decreases (curve II, Fig. 52). It is, however, evident that for angles of α lying between 70° and 90° the change in P is not large and over this range the yield pressure is of the order of $3Y$.

If there is appreciable friction between the indenter and the indentation, the flow pattern is modified. The pressure is no longer uniform but increases towards the apex of the indenter and the mean value of the pressure increases. For moderate

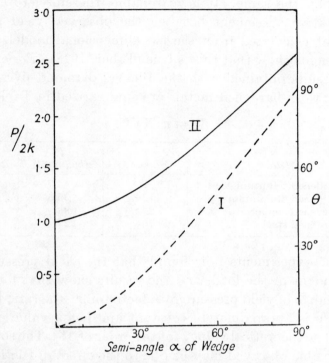

FIG. 52. Penetration of an ideal plastic metal by a two-dimensional wedge of semi-angle α. Curve I, variations of θ (see Fig. 51) as a function of α. Curve II, variation of pressure on wedge as a function of α. (Theoretical results obtained by Hill, Lee, and Tupper, 1947.)

values of μ, however, and for indenters which are not too pointed, the effect does not appear to be very marked.

Indentation by conical and pyramidal indenters

We have already seen in Chapter III that for the two-dimensional flat punch the yield pressure is very similar to that observed for a flat circular punch. We may therefore expect that the solution for the two-dimensional wedge is roughly valid for the three-dimensional indenter, particularly if the semi-angle is not too small. This is supported by the observation that for conical indenters of semi-angle greater than about 60°, for square pyramidal indenters of semi-angle 68°, and for the Knoop indenter where the semi-angle for the two pairs of opposite edges is 86° 15′ and 65°, the hardness value or the yield pressure is essentially the same. It is reasonable, therefore, to expect a quantitative agreement between the observed yield pressure and that calculated from the two-dimensional model. Simple experiments show that this is true (Tabor, 1948). For example, some results obtained with the Vickers diamond pyramid indenting work-hardened metals are summarized in Table XVII.

TABLE XVII

Metal	Y (kg./mm.²)	P (kg./mm.²)	Ratio P/Y
Work-hardened tellurium–lead .	2·1	6·7	3·2
Work-hardened aluminium .	12·3	39·5	3·2
Work-hardened copper . .	27	88	3·3
Work-hardened steel . .	70	227	3·2

In these experiments it is found that the yield pressure P is independent of the load, and the results show that for metals which range in yield pressure by a factor of 30, the ratio between P and Y is approximately constant and has a value of about $3\cdot3Y$. The theoretical value is of the order of $3Y$. The somewhat higher value observed may be due to the effect of friction or to the fact that the two-dimensional model is not as satisfactory for the pointed indenter as for the flat punch. The discrepancy becomes very much more marked for cones or pyramids of small semi-angles. Experiments show that the yield pressure increases as the semi-angle decreases (see full line, Fig. 46). The two-dimensional model, however, provides a decreasing pressure with decreasing angle (see Fig. 52). It would seem that this effect is not entirely attributable to the friction between the indenter and the indentation. Thus even when the frictional effects are reduced to a minimum there is no sign of a decrease in pressure with decreasing angle (broken line, Fig. 46). The discrepancy is probably due to the fact that the flow pattern for three-dimensional deformation is different from that which obtains in two-dimensional deformation. This divergence is not very marked for flat indenters, but for pointed indenters it becomes sufficiently important to yield results that are appreciably in error.

With conical and pyramidal indenters we do not face the problem of the 'onset' of plastic deformation, as with spherical indenters. If the point of the indenter is geometrically perfect it may be considered as a portion of a sphere of vanishingly small radius. Consequently when it first touches the surface it produces plastic flow for the smallest loads. Once penetration has commenced the indentation is sensibly constant in shape and the flow pattern is unchanged whatever the size of the indentation. Consequently the yield pressure does not depend on the size of the indentation, i.e. it is independent of the load.

Indentation of metals which work-harden

It is a relatively simple matter to apply the above conclusions to metals which undergo work-hardening as a result of the indentation process itself. We assume again that there is a representative yield stress Y which is related to the yield pressure P by a relation $P = cY$, where, for the Vickers indenter, c has a value of about 3·3. From the geometry of the pyramid this means that the Vickers hardness number $H_v = cY$, where c has a value of 2·9 to 3. If we compare the stress–strain curve of a metal with its Vickers hardness value at various stages of work-hardening, inspection indicates that the representative deformation produced by the indentation itself corresponds to an additional strain of about 8 to 10 per cent.

The results obtained for steel and copper specimens are shown in Table XVIII (Tabor, 1948). In these experiments Vickers hardness measurements were made on specimens of these metals after they had been deformed by various amounts ϵ_0. The yield stress Y corresponding to a deformation of $(\epsilon_0 + 8)$ per cent. was then determined from the stress–strain curves of the metals. This value is assumed to correspond to the 'representative' value of the yield stress around the indentation. If it is multiplied by a constant c having a value of 2·9 to 3, it should agree with the observed Vickers hardness number. The last two columns show that this is approximately true over a wide range of deformations. We would not expect better agreement over the whole range, since the additional deformation produced by the indentation is not truly additive to any initial deformation. Nevertheless, the agreement indicates that the general picture is reasonably valid. In addition it is clear from the value of the indentational strain (8 per cent.) and from the value of c (2·9 to 3) that the Vickers hardness numbers will be close to the Brinell hardness numbers over an appreciable range of hardness values. The deviation observed with very hard metals has already been discussed in Chapter IV, pp. 55–60.

TABLE XVIII

Metal	Initial deformation (ϵ_0 per cent.)	$\epsilon = (\epsilon_0 + 8)$ (per cent.)	Y at ϵ (kg./mm.^{2})	cY	Observed Vickers hardness number
				$2 \cdot 9Y$	
Mild steel	0	8	55	159	156
	6	14	62	176	177
	10	18	66	190	187
	13	21	67	194	193
	25	33	73	211	209
				$3 \cdot 0Y$	
Annealed copper	0	8	15	45	39
	6	14	20	60	58
	12·5	20·5	23·3	70	69
	17·5	25·5	25	75	76
	25	33	26·6	80	81

Pincushion and barrel-shaped indentations

It is clear from earlier discussions that the detailed shape of the indentation will depend on the degree of work-hardening of the metal under examination. If the metal is highly worked, so that no appreciable work-hardening is produced by the indentation process itself, the metal behaves approximately as an ideally plastic material and the displacement of the metal will follow the flow pattern indicated by the two-dimensional model in Fig. 51. The displaced material will, indeed, tend to flow up the faces of the indenter, but since it will be less constrained on the faces than on the corners, it will rise more at these regions than at the edges. Thus with the Vickers indenter, the indentation will be wider at the centre of the faces than at the edges. This produces, in effect, a barrel-shaped indentation.

If, on the other hand, the metal is annealed, the displaced material is pushed out at some distance from the indenter, as we saw in Chapter V, Fig. 33. The indentation is sunk below the general level of the surface and the effect is more marked at the centre of the faces than at the edges. This in turn produces a pincushion-shaped impression.

Vickers hardness number and the ultimate tensile strength

We may derive a relation between the Vickers hardness number and the ultimate tensile strength in a manner similar to that discussed in Chapter V. If the true stress–strain curve of the metal may be expressed by a relation $Y = b\epsilon^x$, the ultimate tensile strength T_m has the value

$$T_m = b(1-x)\left(\frac{x}{1-x}\right)^x.$$

If the Vickers indentation produces a strain of 8 per cent. the representative yield stress is given by $Y = b(0\cdot08)^x$, so that the Vickers hardness number is given by

$$H_v = 2\cdot9b(0\cdot08)^x.$$

Consequently the ratio of T_m to H_v is given by

$$\frac{T_m}{H_v} = \frac{1-x}{2\cdot9}\left(\frac{12\cdot5x}{1-x}\right)^x. \tag{3}$$

The result obtained is plotted in Fig. 53. It should be noted that here, again, the value of the Meyer index n (where $n \approx x+2$) is still required to determine the value of the ratio (Tabor, 1951).

Rockwell and Monotron hardness

It is not out of place to discuss here the Rockwell and Monotron tests, both of which are based on measurements of the depth of penetration. In the Rockwell test a load of 10 kg. is first applied to the surface and the depth of penetration is reckoned

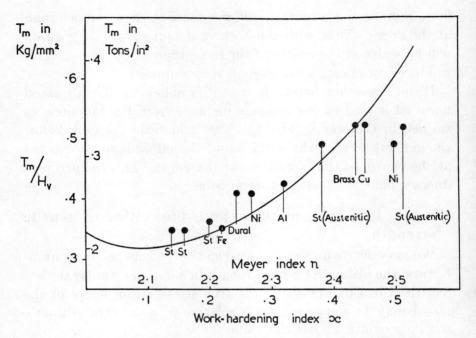

FIG. 53. Ratio of Vickers hardness H_v to ultimate tensile strength T_m as a function of the Meyer index n. ● Brinell results from O'Neill (1934).

as the zero of further measurement. A further load of 90 kg. or 140 kg. is then applied and removed, leaving the minor load, and the additional depth of the indentation recorded directly on a suitable dial gauge. The hardness is then expressed in terms of the dial reading, and the value so obtained may be correlated with Vickers or Brinell hardness values. For softer materials a spherical indenter is used (Rockwell 'B'), for harder metals a conical indenter with a hemispherical tip (Rockwell 'C').

The Rockwell test has two great advantages. First, the application and retention of the minor load prepares the surface upon which the increment in penetration due to the major load is based. Secondly, the hardness value is read directly on the dial gauge and optical measurements of the indentation diameter are unnecessary. As we have seen, however, there may be appreciable elastic 'recovery' when the major load is removed and the depth of the recovered indentation may be considerably less than that of the unrecovered indentation. Consequently hardness values deduced from the depth of the recovered indentation may be appreciably in error. If, of course, the apparatus is calibrated on materials possessing approximately the same Young's moduli, the error may be small. But it is quite possible for two metals to have the same elastic limit but widely different elastic moduli. For example, work-hardened copper and pure iron have elastic limits of about 30 kg./mm.^{2} corresponding to a true yield pressure of about 90 kg./mm.^{2} But the elastic modulus of copper is about 10^{12} dynes/cm.^{2}, whilst for iron it is about 2×10^{12} dynes/cm.^{2} Consequently the recovered indentation for the iron specimen will be shallower than for the copper, so that the penetration method will tend to give a higher hardness value to the iron.

In the Monotron test (Shore Instrument Company) this defect does not apply. Here the load required to produce a fixed depth of penetration is measured. The measurement is made whilst the load is applied so that there are no complications due to elastic recovery. There will, however, be some elastic yielding of the metal during the application of the load, for the indentation process as a whole drags the surface down (elastically) into the metal specimen, as may be seen by comparing Fig. 42 b and Fig. 42 a in Chapter VI. A little consideration shows that this will be of the same order of magnitude as the elastic recovery associated with the shallowing of the indentation when the load is removed, though it will, of course, be in the opposite direction. Thus with the copper and iron specimens discussed in the previous paragraph the load required for the indenter to descend a fixed depth will be less for the copper than for the iron, so that the iron will again appear to be the harder.

Another source of error in the determination of hardness from depth measurements is the effect of 'piling-up' or 'sinking-in' which will vary from one metal to another according to its state of work-hardening. If these effects may be neglected, and if the plastic deformation is large compared with the elastic effects (so that the errors due to elastic recovery or elastic deformation may be considered unimportant), it is relatively simple to derive a relationship between the hardness values obtained from depth measurements and those obtained from direct measurements of the indentation diameter. Suppose the indenter is a sphere of diameter D which undergoes negligible deformation during the indentation process. Suppose it is pressed into the surface with a load W and produces an indentation of chordal diameter $d = 2a$. Then the depth of penetration t is, by simple geometry, given by the relation $(D-t)t = a^2$. Since t is usually small compared with D, we may write

$$t = \frac{a^2}{D} = \frac{d^2}{4D}.$$

Now the mean pressure P across the indentation, which is equivalent to the Meyer hardness, is given by

$$P = \frac{4W}{\pi d^2}.$$

Hence

$$t = \frac{W}{P}\frac{1}{\pi D}. \tag{4}$$

Thus if the depth of penetration is kept constant, as in the Monotron test, the load W is proportional to P. Since the Brinell hardness is only a few per cent. smaller than P (except for deep indentations), this means that W should be approximately proportional to the B.H.N. This is generally found to be the case. For example, the result obtained for the Monotron 'Universal' C–D scale, involving the use of a spherical diamond indenter, is shown in Fig. 54. It is seen that the Monotron hardness value is, in fact, roughly proportional to the Brinell hardness.

If, on the other hand, the load is kept constant, as in the Rockwell test, equation (4) shows that the depth of penetration t is proportional to $1/P$, so that the depth of penetration decreases with increasing hardness. In order to provide an increasing scale with hardness, the Rockwell tester records *not* the actual depth penetrated, but a quantity R given by 100 scale divisions minus the depth penetrated. This gives a scale of hardness values in which the hardness increases with R. We have

$$R = \text{constant} - t.$$

Hence $$R = C_1 - C_2/P, \qquad (5)$$

where C_1 and C_2 are suitable constants. This type of relation is approximately obeyed in the Rockwell test when a spherical indenter is employed. If, however, the conical indenter is employed, a different relation is obtained. If α is the semi-angle

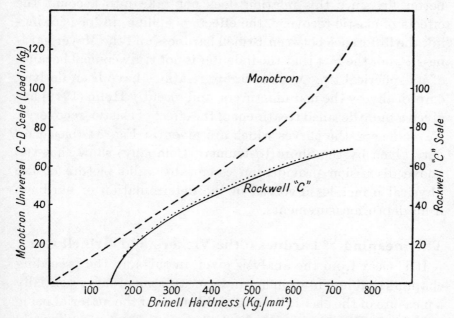

FIG. 54. Relation between Brinell hardness values and hardness measurements based on depth of penetration. For the Monotron test (broken line) the relation is approximately linear; for the Rockwell 'C' test using a spherically tipped cone (full line) the relation is of the parabolic type. These curves are drawn from data issued by the Shore Instrument Company. The dotted line has been drawn to the empirical equation $R_c = 124(1 - 12 \cdot 2/\sqrt{B})$, where B is the Brinell hardness number.

of the cone, the depth is $t = a \cot \alpha$. Since the mean pressure is again given by $P = W/(\pi a^2)$ we have that

$$t = \sqrt{\left(\frac{W}{\pi P}\right)} \cot \alpha. \qquad (6)$$

It follows that the Rockwell number will be expressed by a relation of the type $$R = C_3 - C_4/\sqrt{P}, \qquad (7)$$

where C_3 and C_4 are suitable constants. A typical result showing the relation between the B.H.N. and the Rockwell 'C' hardness (involving the use of a spherically tipped conical indenter) is given in Fig. 54. We again assume that P is not widely different from B, the B.H.N. It is seen that the theoretical relation is approximately obeyed and the curve very roughly fits a relation of the type $R_c = 124(1-12\cdot2/\sqrt{P})$. We should not expect a better fit, since this relation does not take into account the effects of elastic recovery, the effects of 'piling-up' or 'sinking-in', the difference between Brinell hardness and the Meyer hardness P, nor the fact that the indenter is not truly conical because of its spherical tip. More accurate relations have been derived empirically by the manufacturers, and recently Holm (1949) has given a more detailed treatment of the effect of elastic 'recovery'. Nevertheless, the curves which are given in Fig. 54 (based on data given by the Shore Instrument Company) show that the above discussion provides an essentially valid picture of the physical principles involved in the determination of hardness from depth measurements.

The meaning of hardness: the Vickers and Brinell test

It is clear from the analysis given in this and the preceding chapters that indentation hardness measurements are essentially a measure of the elastic limit or yield stress of the material being examined. For most types of indenters in common use the yield pressure between the metal and the indenter when appreciable plastic flow has occurred is about three times the effective yield stress of the metal. Although there is some elastic recovery of the indentation when the indenter is removed, the main change in dimensions occurs in the depth rather than in the projected area of the indentation. Consequently the yield pressure, or the hardness, as determined from the area of the recovered indentation is very nearly equal to the pressure which obtains during the actual formation of the indentation. It is essentially dependent on the plastic properties of the metal and only to a secondary extent on the elastic properties. If, however, the hardness measurements are based on a determination of the depth of the indentation, the elastic recovery may produce marked changes in the calculated pressure, and there may also be appreciable 'piling-up' or 'sinking-in.' In such methods, therefore, the calculated yield pressure may be appreciably different from that which obtains during the actual formation of the indentation.

With pyramidal or conical indenters, where the indentation is geometrically similar whatever its size, the mean pressure to produce plastic flow under the indenter is almost independent of the size of the indentation. Consequently the hardness number has a single value over a wide range of loads. In practice this is very convenient, for it means that with this type of hardness measurement it is not necessary to specify the load. However, this test deprives us of a means of estimating the degree of work-hardening of the metal.

With spherical indenters, on the other hand, the shape of the indentation varies with its size, so that the amount of work-hardening and hence the elastic limit increases with the size of the indentation. As a result the yield pressure in general increases with the load. For this reason, in Brinell hardness measurements it is necessary to specify the load and the diameter of the indenter. However, this increase in yield pressure with size of indentation provides useful information about the metal; first, about the elastic limit or yield stress of the specimen, and secondly, about the way in which the yield stress increases with the amount of deformation. For this reason, although such measurements involve more trouble than measurements with conical or pyramidal indenters, they provide more information. Indeed, hardness measurements made with spherical indenters and the Meyer analysis to which they may be subjected enable us to determine the degree of work-hardening of the metal under examination and to deduce its stress–strain characteristics.

REFERENCES

Bishop, R. F., Hill, R., and Mott, N. F. (1945), *Proc. Phys. Soc.* (*Lond.*) **57**, 147.

Hankins, G. A. (1925), *Proc. Instn. Mech. Engrs.* **1**, 611.

Hill, R., Lee, E. H., and Tupper, S. J. (1947), *Proc. Roy. Soc.* A **188**, 273.

Holm, E., Holm, R., and Shobert (II), E. I. (1949), *J. App. Phys.* **20**, 319.

KNOOP, F., PETERS, C. G., and EMERSON, W. B. (1939), *Nat. Bureau of Standards*, **23** (1), 39.

LEA, F. C. (1936), *Hardness of Metals*, Charles Griffin & Co., London.

LUDWIK, P. (1908), *Die Kegelprobe*, J. Springer, Berlin.

LYSAGHT, V. E. (1946) *Amer. Soc. Test. Mat. Bulletin No.* 138, 39–44.

ROCKWELL, S. R. (1922), *Trans. Amer. Soc. for Steel Treating*, **2**, 1013.

SMITH, R., and SANDLAND, G. (1922), *Proc. Instn. Mech. Engrs.* **1**, 623.

TABOR, D. (1948), *Proc. Roy. Soc.* A **192**, 247; *Engineering* (1948), 165, 289.

—— (1951), *J. Inst. of Metals*.

WILLIAMS, S. R. (1942), *Hardness and Hardness Measurements*, Amer. Soc. Met.

CHAPTER VIII

DYNAMIC OR REBOUND HARDNESS

Indentation produced by impact

THE dynamic hardness of a metal may be defined, by analogy with static hardness, as the resistance of the metal to local indentation when the indentation is produced by a rapidly moving indenter. In most practical methods the indenter is allowed to fall under gravity on to the metal surface. It rebounds to a certain height and leaves an indentation in the surface. Martel showed in 1895 that, over a wide range of experimental conditions, the volume of the indentation so formed is directly proportional to the kinetic energy of the indenter and these observations have been confirmed by Vincent (1900) and other workers. It is easy to show that this implies that the metal offers an average pressure of resistance to the indenter equal numerically to the ratio

$$\frac{\text{energy of indenter}}{\text{volume of indentation}}.$$

For example, suppose the indenter is of spherical or conical form and suppose that there is a constant dynamic pressure P resisting indentation during impact. If at any instant the projected area of the indentation is A, the force exerted by the metal on the indenter is PA. If in the next instant the indenter penetrates a further distance dx, the work done will be $PA\,dx$. The total work expended in forming the indentation is simply

$$\int PA\,dx = P\int A\,dx = PV,$$

where V is the volume of the indentation. According to Martel this may be equated to the energy of impact, so that

$$P = \frac{\text{energy of impact}}{V}.$$

This has the dimensions of pressure and is sometimes referred to as the dynamic hardness number. Later workers have discussed the validity of this relation in some detail. In particular it has been suggested that the energy of rebound should be taken into account in calculating the dynamic hardness. It is, indeed, evident that the energy expended in producing the indentation, after rebound has occurred, is equal to the energy of impact minus the energy of rebound. This quantity is then equal to the product of P and the volume of the indentation remaining in the surface. It would appear that in the earlier work confusion arose from the fact that there was no clear distinction between the indentation formed during the collision and the indentation remaining after rebound had occurred. As we shall see below, there may be an appreciable change in the volume of the indentation (as a result of elastic recovery) when rebound occurs, and this change as well as the energy of rebound must be taken into account in calculating the dynamic hardness.

A different approach is that adopted by Shore (1918) and by Roudié (1930). In these methods the height of rebound itself is used as a measure of the dynamic hardness. It is found that if the height of fall is constant, the height of rebound is roughly proportional to the static hardness of the material concerned. In what follows we shall analyse the processes involved in impact experiments and develop a simple theory which explains a number of empirical relations observed in dynamic hardness measurements.

Four main stages of impact

Suppose a hard spherical indenter is dropped on to the horizontal flat surface of a softer metal which is in the form of a massive anvil. The indenter strikes the surface and rebounds, leaving an indentation in the surface. The impact may be divided into four main stages. At first the region of contact will be deformed elastically, and if the impact is sufficiently gentle the surfaces will then recover elastically and separate without residual deformation. The collision in this case is purely elastic and the time of impact, mean pressures, and deformations are given by Hertz's equations for elastic collisions. The second stage occurs if the impact is such that the mean pressure exceeds about $1 \cdot 1Y$, where Y is the yield stress or elastic limit of the metal. A slight amount of plastic deformation will occur and the collision will no longer be truly elastic. As we shall see below, this onset of plastic deformation occurs for extremely small impact energies. At higher energies of impact the deformation rapidly passes over to a condition of 'full' plasticity (stage 3) and full-scale plastic deformation proceeds until the whole of the kinetic energy of the indenter is consumed. Finally, a release of elastic stresses in the indenter and in the indentation takes place as a result of which rebound occurs (stage 4).

A full analysis of the four phases involved in the collision process is extremely complicated and difficult. An attempt has been made by Andrews (1930), who was mainly concerned with the *time* of collision, to calculate the time involved in each part of the collision process, but the treatment was of an admittedly approximate nature. If, however, we restrict ourselves to a consideration of the forces involved, and not the time of collision, we may simplify the analysis considerably (Tabor, 1948).

Mean dynamic yield pressure P

We assume that there is a dynamic yield pressure P which to a first approximation is constant and which is not necessarily the same as the static pressure necessary to cause plastic flow. This assumption implies that whenever the pressure during impact reaches the value P, plastic flow occurs and so long as plastic flow continues the pressure remains constant at this value. If, after impact has occurred, the volume of the remaining permanent indentation is V_r, the work done as plastic energy in producing this indentation is, by definition of P, given by

$$W_3 = PV_r. \tag{1}$$

Clearly the energy W_3 is the difference between the energy of impact W_1 and the energy of rebound W_2. All that remains, therefore, is to calculate W_2 and the volume V_r.

Suppose the indenter has a mass m and a spherical tip of radius of curvature r_1 and that it falls from a height h_1 on to a flat metal surface. After the collision the indenter rebounds to a height h_2 and leaves a permanent indentation in the metal surface of chordal diameter $d = 2a$ (Fig. 55). We assume that the mechanism involved in the dynamic indentation is essentially the same as that which occurs under static conditions.

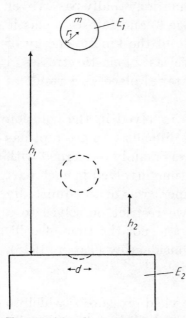

That is to say, when the plastic deformation has been completed there is a release of elastic stresses in the indenter and in the indentation. We assume that the energy involved in the release of these elastic stresses is equal to the energy of rebound of the indenter. Finally, we assume that Young's moduli for the indenter and the metal are essentially the same as for static conditions.

Consider the indentation after the impact has occurred. Since there has been a release of elastic stresses in the indentation, its radius of curvature will not be r_1 but will be somewhat greater, say r_2. If we were to apply a suitable load F to the indenter for a very short interval, it would deform the indentation (and itself) elastically,

FIG. 55. A hard spherical indenter of mass m, radius r_1, falls from a height h_1 on to a massive anvil, rebounds to a height h_2 and leaves an indentation of chordal diameter d.

and according to Hertz's equation it would just touch over the diameter d, where

$$d = 2a = \left[\frac{6Fr_1r_2}{r_2-r_1} f(E) \right]^{\frac{1}{3}}, \qquad (2)$$

where $f(E) = (1-\sigma_1^2)/E_1 + (1-\sigma_2^2)/E_2$. Here E_1 and E_2 are Young's moduli for the indenter and the surface respectively, and σ_1, σ_2 are Poisson's ratios.

We estimate the elastic energy involved in this process by calculating the external work performed in pressing the indenter into the indentation. As the indenter sinks into the indentation the force increases from zero and reaches the final value F given by equation (2) as the full contact across the diameter $d = 2a$ is completed. At any intermediate instant when the region of contact has a diameter 2α (where $\alpha < a$) the force $\mathscr{F}$ on the indenter, given by equation (2), is

$$\mathscr{F} = F \frac{\alpha^3}{a^3}. \tag{3}$$

At this stage, as a result of the elastic deformation of both contacting surfaces, the centre of the indenter has descended a distance z (Prescott, 1927) given by

$$z = \frac{3\mathscr{F}}{4\alpha} f(E), \tag{4}$$

where $f(E)$ has the same meanings as in equation (2). Substituting in equation (4) from equation (3) we have

$$z = \frac{3F\alpha^2}{4a^3} f(E). \tag{4 a}$$

Then the integral of $\mathscr{F} \, dz$ over the range $\alpha = 0$ to $\alpha = a$ is the total elastic energy K stored in the surfaces. We have

$$K = \int \mathscr{F} dz = \int_0^a \frac{3}{2} \frac{F^2}{a^6} f(E) \alpha^4 \, d\alpha$$

$$= \frac{3}{10} \frac{F^2}{a} f(E). \tag{5}$$

But this process is the exact converse of what happens when the surfaces recover elastically and the indenter is ejected from the indentation. Since the process is elastic, the energy involved in both cases is the same, so that K is equal to the energy of rebound.

Hence

$$W_2 = mgh_2 = \frac{3}{10} \frac{F^2}{a} f(E). \tag{6}$$

The volume V_r of the permanent indentation left in the surface may be written to a first approximation as $V_r = (\pi a^4)/(4r_2)$.

Hence

$$W_3 = W_1 - W_2 = PV_r = P \frac{\pi a^4}{4r_2}. \tag{7}$$

We now express r_2 in terms of r_1 and F from equation (2).

$$\frac{1}{r_2} = \frac{1}{r_1} - \frac{3}{4}\frac{F}{a^3}f(E).$$

Hence
$$W_3 = P\frac{\pi a^4}{4r_1} - \frac{3}{16}\frac{F^2}{a}f(E), \qquad (8)$$

since the force F at the end of indentation is equal to $P\pi a^2$. The first term of equation (8) is simply PV_a, where V_a is the *apparent* volume of the indentation which would be obtained if the indentation were considered to have the same radius of curvature as the indenter. The second term, by comparison with equation (6), is seen to be equal to $\frac{5}{8}W_2$.

Hence
$$W_3 = PV_a - \tfrac{5}{8}W_2. \qquad (9)$$

Thus
$$P = \frac{mg(h_1 - \tfrac{3}{8}h_2)}{V_a}. \qquad (10)$$

The validity of this analysis depends on the assumption that the internal forces occurring in the actual impact are essentially the same as those involved in the analytical model just described. In particular, we assume that the elastic waves set up in the indenter and the metal specimen absorb a negligible amount of energy. We also assume that the temperature rise of the material around the indentation during the impact is small and has a negligible effect on the strength properties of the metal.

It is at once apparent that if the rebound is not very large (so that h_2 is small) the results will not be very different from the equation given by Martel, $P = mgh_1/V_a$, nor from the equation suggested by later workers, $P = mg(h_1 - h_2)/V_a$.

The effect of variation in the value of P

In the above derivation we have assumed that P is a constant throughout the process of impact. There are, however, two reasons why we may expect P to vary during the collision. The first is a dynamic effect associated with the kinetic displacement of the metal during impact. This will tend to increase P at the initial stages of the deformation when the velocity of displacement is a maximum (see later). It is difficult, however, to express this effect quantitatively. The second reason is that work-hardening of the deformed material will occur during the formation of the indentation. As a result, P will tend to increase during impact in a manner similar to that observed in static hardness measurements, as described in Chapter II. We may make some estimate of the order of this effect by assuming that, on analogy with the static indentations, we can write

$$P = kd^{n-2},$$

where n lies between 2 and 2·5. Then the work W_3 expended in displacing plastically a volume V_r becomes

$$W_3 = \frac{4}{n+2} PV_r, \tag{1a}$$

where P is now the mean pressure at the *end* of the deforming process. This is also the pressure involved in the calculation of the rebound. Substituting this value of W_3 in the appropriate equations, we obtain for the mean pressure at the *end* of the indentation process,

$$P = \frac{n+2}{4} \frac{mg\left(h_1 - \frac{2n-1}{2(n+2)} h_2\right)}{V_a}. \tag{10a}$$

The last term in the bracket varies from $\frac{3}{8}h_2$ to $\frac{4}{9}h_2$ as n varies from 2 to 2·5, so that this term tends to give lower values for P. On the other hand, the term in front of the main bracket increases from 1 to 1·12 as n increases from 2 to 2·5. The total effect is to give values of P which are somewhat greater than those given in equation (10). The difference, however, will never be more than about 10 per cent.

The validity of equations (6) and (10)

Remembering that at the end of the indentation process, $F = \pi a^2 P$, we rewrite equation (6) as

$$h_2 = \frac{3}{10} \frac{\pi^2 a^3 P^2}{mg} \left(\frac{1-\sigma_1^2}{E_1} + \frac{1-\sigma_2^2}{E_2}\right). \tag{6a}$$

Assuming a value of 0·3 for Poisson's ratio for both surfaces, this gives

$$h_2 = \frac{2\cdot7 a^3 P^2}{mg} \left(\frac{1}{E_1} + \frac{1}{E_2}\right). \tag{6b}$$

Since the apparent volume of the indentation V_a is proportional to a^4, this means that h_2 is proportional to $V_a^{\frac{3}{4}}$ for any fixed material. Plotting h_2 against V_a on logarithmic ordinates, we should obtain straight lines with a slope of $\frac{3}{4}$, if P is constant. Some results taken from Edwards and Austin's paper (1923) are plotted in Fig. 56, and it is seen that this is approximately true.

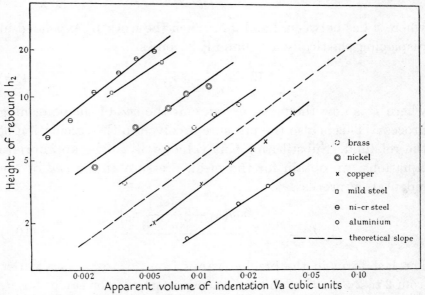

FIG. 56. Plot of height of rebound (h_2) against apparent volume of indentation (V_a) for a spherical indenter impacting on various metals. The experimental results are from a paper by Edwards and Austin (1923) and it is seen that for each metal the points lie on a straight line of slope close to the theoretical value.

If P is not constant but varies in the manner given by equation (11) we find that the logarithmic graph of h_2 against V_a is still a straight line, but the slope has a value of $(3+2n-4)/4$, i.e. it varies from $\frac{3}{4}$ to 1 as n varies from 2 to 2·5. It is seen from Fig. 56 that in fact the points for each material lie on a straight line, and that the slope lies between 0·7 and 0·85.

Again, for indentations of a fixed diameter, h_2 should be proportional to $P^2\left(\dfrac{1}{E_1}+\dfrac{1}{E_2}\right)$. If, therefore, we plot h_2 against $P\sqrt{\left(\dfrac{1}{E_1}+\dfrac{1}{E_2}\right)}$ on logarithmic ordinates, we should obtain a straight line of slope $\frac{1}{2}$. Results taken from the same paper are plotted in Fig. 57, the values of P being calculated according to equation (10). There is again good agreement, the slope of the straight line being 0·51. In this case, any dependence of P on the size of the indentation does not appreciably affect the relation.

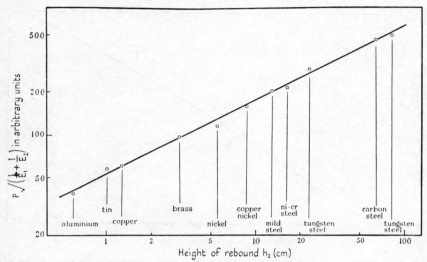

FIG. 57. Plot of $P\sqrt{(1/E_1+1/E_2)}$ against h_2 for a fixed size of indentation. The results are based on data given by Edwards and Austin (1923) and the points lie on a straight line of slope 0·51. The theoretical slope is 0·5.

Finally, we may eliminate a between equations (6 a) and (10). The resulting relation between h_1, h_2, and P is given by

$$P^5 = \frac{h_2^4}{(h_1-\frac{3}{8}h_2)^3} \frac{mg}{r_1^3} \frac{10^4}{\pi^5 4^3 3^4} \left(\frac{1}{f(E)}\right)^4. \qquad (11)$$

A similar relation is obtained if equation (10 a) is used instead of equation (10). If we assume a value of 0·3 for σ_1 and σ_2, equation (11) gives

$$P^5 = \frac{h_2^4}{(h_1-\frac{3}{8}h_2)^3} \frac{mg}{109 r_1^3} \frac{1}{(1/E_1+1/E_2)^4}. \qquad (11a)$$

Since the bracket involving Young's moduli does not vary greatly for most metals, we may treat this factor as a constant and plot P as a function of h_2 for a given height of fall h_1. The theoretical curve is shown in Fig. 58. If we allow for the fact

that softer metals usually have a smaller Young's modulus, the
curve is modified in a manner similar to that shown in the dotted
curve. This analysis assumes that P is constant. If there is any
increase in P due to work-hardening, it may be allowed for
approximately by using equations $(6\,a)$ and $(10\,a)$, but the resulting curve is not very different from that given in Fig. 58. The

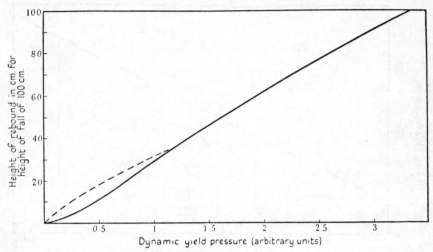

FIG. 58. Dynamic yield pressure (or hardness) as a function of the height of
rebound for a fixed height of fall of 100 cm. Full line—assuming that Young's
modulus is the same for all metals; broken line—assuming that for softer metals
Young's modulus has a smaller value than for hard metals.

effect of the velocity of collision on P is more difficult to allow
for, but in any case the results given in Fig. 58 will give a fairly
reliable value of the pressure developed in the course of the
collision process.

It is, however, clear that even if the metal does not work-
harden, and even if the velocity of impact has no effect on P,
the dynamic yield pressure given by Fig. 58 will not be a single-
valued constant of the metal. It will depend on the size of the
indentation formed, corresponding to the transition from the
onset of plastic deformation to full plasticity. Thus with a light
impact, where the collision may be almost entirely elastic, the
pressure P involved will be only about $1 \cdot 1/2 \cdot 8 = 1/2 \cdot 5$ times the
pressure involved in a heavier impact where appreciable deforma-
tion occurs. This has been observed in some impact experiments

described by Davies (1949). With a steel ball of diameter 1 cm. falling on to a tool steel surface, the onset of plastic deformation occurred for a height of fall of 1 cm. corresponding to a yield pressure P of about 160 kg./mm.^{2} For a large indentation, however, the yield pressure was of the order of 360 kg./mm.^{2}

It is evident, therefore, that Fig. 58 is approximately valid, if the ordinate refers to the yield pressure during the collision itself. If by yield pressure we mean the pressure associated with 'full' plastic deformation the results in Fig. 58 will be valid for the smaller heights of rebound since here the impressions will be fairly large. It will not be valid for the greater heights of rebound since here the collision process approaches the conditions associated with the onset of plastic deformation. In this region, as we have seen, the yield pressure associated with full plasticity will be two to three times larger. Thus if we wish to plot height of rebound (for a fixed height of fall) as a function of the yield pressure at the fully plastic stage, the lower portion of the curve will be similar to that given in Fig. 58, whilst the upper portion will be pushed over to the right. This will introduce a more marked S-shaped curve into the characteristic. This discussion is relevant in the calibration of the Shore rebound scleroscope.

Shore rebound scleroscope

In the Shore rebound scleroscope a small indenter supported in a vertical glass tube falls from a height of about 25 cm. on to the specimen to be tested and the height of rebound is observed. The height of fall is divided into 140 equal divisions and hard steels give a height of rebound of the order of 90 to 100 divisions. The tip of the indenter may be a diamond with a spherical end or a hard steel ball of diameter 3 mm. The calibration characteristic is shown in Fig. 59 (Shore, 1918), where the height of rebound is plotted against the Brinell hardness of the metal. It is seen that there is a rather wide scatter in experimental results. Nevertheless, the scleroscope is often useful in determining the hardness of massive engineering parts since the rebound measurements may be carried out *in situ*.

As we shall see later, for most metals at a velocity of impact corresponding to a height of fall of 25 cm. the dynamic yield pressure is not widely different from the static yield pressure. Consequently we should expect agreement between the calibration curve in Fig. 59 and the theoretical curve in Fig. 58. However, the B.H.N.s are obtained when appreciable plastic deformation has occurred, whereas for the harder metals the

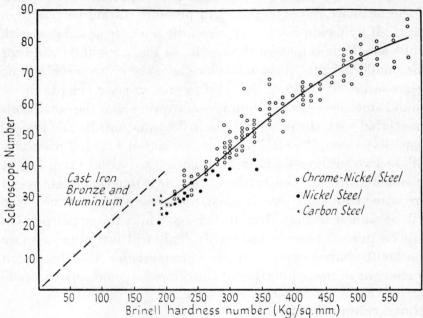

Fig. 59. Calibration characteristic of the Shore rebound scleroscope, showing the height of rebound (for a fixed height of fall) as a function of the static hardness of metals (from data given by Shore, 1918). The curve is similar to the theoretical curve given in Fig. 58.

scleroscope produces relatively small indentations. Consequently the results of Fig. 58 should be modified in the light of the discussion given above and should show a more pronounced **S**-shaped curve for the harder metals. This is apparent in Fig. 59, and the general calibration curve of the sclerometer is in reasonably good agreement with the theoretical results.

The condition for elastic collisions

It is interesting to consider what happens when the rebound is equal to the height of fall. In this case the processes of impact and rebound become entirely elastic, and there is no plastic deformation of the anvil. If we go back to the Hertzian equations and calculate the final average pressure P_e developed between the indenter and the anvil in a purely elastic collision, we obtain a relation

$$P_e^5 = \frac{8^3}{5^3} \frac{10^4}{\pi^5 4^3 3^4} \frac{mgh_1}{r_1^3} \left(\frac{1}{f(E)}\right)^4, \tag{12}$$

or putting $\sigma_1 = \sigma_2 = 0.3$

$$P_e^5 = \frac{1}{26 \cdot 6} \frac{mgh_1}{r_1^3 (1/E_1 + 1/E_2)^4}. \tag{12a}$$

But equations (12) and (12 a) are exactly the same as the value for P obtained from equations (11) and (11 a) when h_2 is put equal to h_1. Two conclusions follows from this result. Firstly, equation (11) is valid right up to a rebound of 100 per cent. In the latter case, the pressure obtained from equation (12) is then the final mean pressure between indenter and specimen. Secondly, plastic deformation will not occur if the yield pressure of the specimen is higher than the value of P_e given in equation (12). If it is less than P_e, plastic deformation occurs and the value of P in equation (11) gives the dynamic yield pressure of the material. This approach has been used by Davies (1949) to study the onset of plastic yielding under dynamic conditions (see also Taylor, 1946). A similar approach has also been made by Tagg (1947) in discussing the deformation of steel pivots in sapphire jewels as a result of impact. It is instructive to carry out a few calculations and to compare the conditions for plastic yielding under static conditions with those occurring under impact.

Suppose the indenter in both the static and dynamic experiments is a steel ball of diameter 1 cm. (mass 4 g.). We may calculate the static load required to produce the onset of plastic deformation ($P = 1 \cdot 1Y$) as described in Chapter IV. The results are given in the fourth column of Table XIX. We may now calculate the height from which the ball must be dropped to produce the onset of plastic deformation under impact. We use the value of P_e given in equation (12 a) and assume that plastic yielding again commences when $P_e = 1 \cdot 1Y$. This is not accurately true, but it is a good approximation. The results are given in the fifth column of Table XIX.

TABLE XIX

Conditions Necessary to produce Onset of Plastic Deformation under Static and Dynamic Conditions (Diameter of ball = 1 cm.)

Metal	Approximate Brinell hardness (kg./mm.²)	Static yield stress Y (kg./mm.²)	Static load at which onset of plastic deformation occurs (g.)	Height from which ball must be dropped to produce onset of plastic deformation (cm.)
Tellurium–lead . .	6	2·1	2	$0·5 \times 10^{-6}$
Soft copper . .	55	20	62	$3·2 \times 10^{-4}$
Work-hardened copper	90	31	230	$2·8 \times 10^{-3}$
Work-hardened mild steel . . .	190	65	1,200	$1·5 \times 10^{-2}$
Alloy steel . . .	350	130	9,500	0·5

It is at once apparent from Table XIX that extremely light impacts are sufficient to produce plastic deformation of metal surfaces. Thus even with a hard alloy-steel, where a static load of 9,500 g. is necessary, a height of fall of less than 1 cm. is all that is required to produce the onset of plastic deformation. We may note that the ball weighs only 4 g. The reason for this marked sensitivity to impact is that the time of collision is very short (see below) so that the impulsive forces developed, even for light impacts, may be very high indeed.

It is also evident that the rebound method of determining the dynamic hardness will not discriminate between materials possessing yield stresses above P_e since they will all give a rebound of 100 per cent.† To increase the range of the method the experimental conditions must be modified to give a higher value of P_e in equation (12). This may be readily achieved by increasing the mass of the indenter or the height of fall. For example, an increase of either of these by a factor of 32 will double the value of P_e. A more sensitive method, however, is to

---

† In practice, of course, 100 per cent. rebound is never attainable since some energy is lost by the dissipation of elastic waves within the bodies, and by elastic hysteresis losses if the elastic relaxation time is large compared with the time of collision.

decrease the radius of the tip of the indenter. A decrease of r by a factor of 3·2 will double the value of P_e. These observations may be of value in the design of impact hardness equipment.

Coefficient of restitution

If the indenter falls with a velocity v_1 on to the surface of the anvil and rebounds with a velocity v_2, the coefficient of restitution e is defined as
$$e = v_2/v_1.$$

We may find e from equation (11) by putting $v_1^2 = 2gh_1$, $v_2^2 = 2gh_2$. Assuming that the yield pressure P remains essentially constant, equation (11) gives

$$v_2 = k(v_1^2 - \tfrac{3}{8}v_2^2)^{\tfrac{3}{4}}. \tag{13}$$

It is clear from this relation that v_2 does not vary linearly with v_1 so that the ratio $e = v_2/v_1$ will not be a constant. The way in which e varies with the velocity of impact is shown in Fig. 60, where curves i, ii, iii, iv, and v respectively have been drawn for values, at an impact velocity of 450 cm./sec., of $e = 1$, 0·8, 0·6, 0·4, and 0·2. (This velocity corresponds approximately to a height of fall of 100 cm.) As we shall see later, P is not a constant, so that we may expect some deviation from these curves in practice. Nevertheless, the general form of these curves is fully confirmed in practical experiments. Typical results obtained for cast steel and drawn brass are shown in the dotted lines. Similar curves have been obtained by Raman (1918), Okubo (1922), and Andrews (1930) in experiments on the impact of spheres of similar metals. Although in this case *both* spheres are plastically deformed at the region of contact, the relation between v_1 and v_2 is of the same type as in equation (13).

If instead of equation (10) we use equation (10 a) to derive the relation between v_1 and v_2, we obtain

$$v_2 = k(v_1^2 - \beta v_2^2)^{\beta}, \tag{13 a}$$

where $\beta = (2n-1)/(2n+4)$. This equation gives curves which are similar to those given by equation (13), but they are appreciably steeper near the origin and flatter beyond.

It is apparent from these equations and from the experimental curves that in general the coefficient of restitution of impacting solids capable of undergoing plastic deformation will not be a constant. At very low velocities of impact the pressures developed will be insufficient to cause plastic flow. The collision process will be entirely elastic and the coefficient of restitution will be unity (except for small elastic hysteresis losses). This occurs even with the softest metals if the velocity of impact is

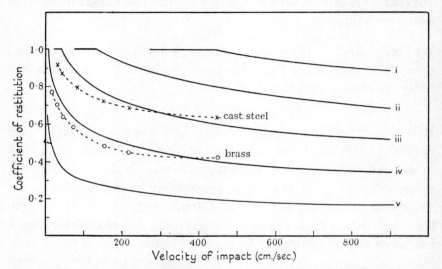

FIG. 60. Variation of the coefficient of restitution with the velocity of impact for materials of various hardness. Full lines—theoretical curves; dotted lines—experimental curves for cast steel and drawn brass.

small enough, as Andrews (1931) showed for lead and tin alloys. As the velocity of impact increases, the amount of plastic deformation will steadily increase, and there will be a corresponding decrease in the coefficient of restitution.

The time of impact: elastic collisions

A full calculation of the time of impact of colliding metals is extremely complicated and has not yet been satisfactorily solved. If the bodies are very long the period of impact may be determined by the time for an elastic compression wave to travel along the bodies and be reflected back to the colliding surface, as St.-Venant showed many years ago. If, however, the bodies are relatively short the time of impact is determined predominantly by the processes which occur at the actual region of impact. Consequently if the collision involves only elastic deformations the time of impact is given accurately by Hertz's equations. If the maximum distance the indenter sinks into the surface during the collision is z_0 and the velocity of impact is v, then the total time of impact t_e according to the elastic equations (Hertz, 1881) is

$$t_e = 2 \cdot 94 \frac{z_0}{v}.$$

The value of z_0 is obtained from equation $(4a)$ by putting $\alpha = a$. We have

$$z_0 = \frac{3}{4} \frac{F}{a} f(E) = \frac{3\pi Pa}{4} f(E). \qquad (4b)$$

Eliminating P and a by using equations (6) and (12) we may find z_0 and hence determine t_e. The result, assuming

$$\sigma_1 = \sigma_2 = 0 \cdot 3,$$

is

$$t_e = \frac{2 \cdot 74 m^{\frac{2}{5}}}{v^{\frac{1}{5}} r^{\frac{1}{2}}} \left(\frac{1}{E_1} + \frac{1}{E_2} \right)^{\frac{2}{5}}. \qquad (14)$$

It is seen from this equation that the time is larger for small velocities of impact but the rate of variation with v is extremely slow. We may as a matter of interest calculate a typical impact time. Suppose the indenter is a steel sphere of radius $r = 0 \cdot 5$ cm., mass $m = 4$ g., and it falls on to an alloy-steel block of yield stress 130 kg./mm.[2] As we saw in Table XIX, the indenter can fall from a height of $0 \cdot 5$ cm. and still give an elastic collision. The duration of the impact for this height of fall $(v = 31$ cm./sec.) may be calculated from equation (14) and is found to be about 5×10^{-5} sec. If we think of the impact as an instantaneous change in momentum, the mean force F_m during the impact is given by

$$F_m t_e = 2mv,$$

so that F_m is about 5 kg. Actually the impulsive force rises from zero to a value of $9 \cdot 5$ kg. and as we saw above a static load of this amount will just cause plastic yielding. These calculations indicate that the high local pressures developed during elastic impact are due to the very short duration of the collision. It is for this reason that plastic deformation is so easily initiated by impact.

Time of impact: plastic collisions

We may now consider the more general case in which the stresses during collision exceed the elastic limit and plastic flow occurs. The process, as we saw above, involves four main parts, and it is not possible to calculate accurately the time involved in each part of the collision process. An analysis has been given by Andrews (1930), but it is of an admittedly approximate nature. As we are mainly concerned with the order of magnitude we may make a crude initial assumption which leads to a very simple solution.

We assume that the indenter is a sphere of radius r, mass m, and that it is completely undeformable. Secondly we consider impacts for which the collision process is predominantly plastic so that elastic strains may be neglected. We assume that the yield pressure has an average constant value P.

At any instant when the sphere has penetrated a distance x the indentation has a radius a, where to a first approximation $2rx = a^2$. The decelerating force on the sphere is equal to $P\pi a^2$ or $P\pi 2rx$. Hence the equation of motion becomes

$$P\pi 2rx = -m\frac{d^2x}{dt^2},$$

or

$$\frac{d^2x}{dt^2} + \frac{2\pi rP}{m}x = 0.$$

The solution is $x = A \sin\sqrt{\left(\frac{2\pi rP}{m}\right)}t$. The sphere is brought to rest when $dx/dt = 0$, i.e.

$$t = \frac{\pi}{2}\sqrt{\left(\frac{m}{2\pi Pr}\right)}. \tag{15}$$

A similar equation was derived by Andrews (1930). It is seen that *the time of collision is independent of the velocity of impact*. For a hard steel ball of diameter 1 cm. ($m = 4$ g.) the time of collision with anvils of various metals is given in Table XX.

It is seen that the time of collision is of the order of several 10^{-5} sec., so that for an indenter of this size it is of the same order of magnitude as the duration of an elastic collision. A comparison of equations (14) and (15) shows that this similarity

TABLE XX

Time of Impact of Hard Sphere, Diameter 1 cm., Mass 4 g.,
striking a Massive Anvil of a Softer Metal. Plastic Collisions

Metal	P (kg./mm.^{2})	t (sec.)
Tellurium–lead . . .	6	7×10^{-5}
Soft copper	55	3×10^{-5}
Work-hardened copper . .	90	2×10^{-5}
Work-hardened mild steel .	190	$1 \cdot 3 \times 10^{-5}$

in the value of t will be quite general but that at very small velocities of impact the elastic collision time will be appreciably longer than the plastic collision time.

Electrical measurement of the time of impact

Some simple experiments carried out in 1939 demonstrate the validity of equation (15) and show the extreme ease with which plastic flow may be initiated during impact (Bowden and Tabor, 1941). In these experiments the collision was observed between suspended spheres of equal radii and the electrical conductance across the spheres recorded on a cathode-ray oscillograph during the impact. The conductance is a measure of the area of contact between the spheres.

Typical results for spheres of radius 2 cm. are shown in Fig. 61 a, b, and c. One sphere is stationary and the other sphere strikes it with a velocity determined by the amplitude of the swing. Fig. 61 a is a record obtained by Dr. Hirst using very hard steel spheres and a low velocity of impact (10 cm./sec.). It is seen that the conductance curve is symmetrical, indicating that the collision is essentially elastic. The elastic equations indicate that the collision time should be of the order of 10^{-3} sec., and this is close to the observed value. At higher velocities of impact, however, the results are similar to those shown in Figs. 61 b and 61 c and the conductance trace is markedly asymmetric.

Fig. 61 b is for mild steel spheres and Fig. 61 c for lead spheres impinging at a velocity of 76 cm./sec. Contact commences at the point A and the curve AC corresponds to the period during which elastic deformation and the initial onset of plastic deformation take place. The curve CD corresponds to the period during which full plastic deformation occurs, whilst DEB represents the separation of the surfaces under the released elastic stresses. The point D is not defined very sharply, so that the period from A to D cannot be estimated to an accuracy better

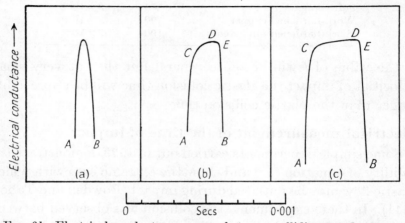

FIG. 61. Electrical conductance curves between colliding metal spheres: (a) tool steel colliding at 10 cm./sec. The curve is symmetrical indicating essentially elastic conditions, (b) mild steel colliding at 76 cm./sec, (c) lead spheres colliding at 76 cm./sec. The asymmetric curves in (b) and (c) indicate that plastic deformation has occurred.

than a few per cent. This error is, however, not important compared with the experimental variation from trace to trace. If we assume that the collision from A to D corresponds to a deformation process which is essentially plastic, we may calculate the period by a method similar to that used in obtaining equation (15).

Instead of giving the result at once it is interesting to consider this impact problem in somewhat greater detail. The collision between a stationary sphere and a sphere of equal size travelling at a velocity V along the line of centres is equivalent to the impact occurring between the spheres when they collide at equal and opposite velocities $V/2$. (For complete equivalence a constant velocity $V/2$ must be superimposed on both spheres.) The plane of contact in this 'equivalent' collision may thus be considered as the stationary plane during the deceleration of the spheres. As before, we consider a predominantly plastic collision. Then at any instant when each sphere has formed a flattened cap of depth x, and chordal radius a, we may to a first approximation write $2rx = a^2$, where r is the radius of the spheres. The decelerating force on each sphere is equal to $P\pi a^2$ or $P\pi 2rx$. Since the spheres have been decelerated over a distance x, the equation of motion of each sphere becomes

$$P\pi 2rx = -m\frac{d^2x}{dt^2}.$$

Both spheres are brought to rest, i.e. the spheres are at their closest point of approach when $dx/dt = 0$. As before (see equation (15)), this occurs after a time

$$t_{AD} = \frac{\pi}{2}\sqrt{\left(\frac{m}{2\pi Pr}\right)}. \qquad (15a)$$

If the spheres were of a material which exhibits no elastic recovery the collision process would come to an end at this stage, but if they possess some elastic properties they will still remain in contact for a further interval of time as they separate under the influence of the released elastic stresses.

The main defect in this analysis is the assumption that the distance over which the spheres are decelerated during the collision is equal to the depth of the flattened cap at the region of contact. In general, apart from the flattening at the region of contact there is an overall *elastic* compression of the spheres as a whole during their approach. Andrews (1930) suggested that this may be allowed for in the following way. If the flattening of the spheres were elastic (and not plastic), the overall elastic compression of the sphere would, according to the Hertzian equation, be exactly equal to the flattening of the cap itself (see, for example, Prescott (1927), p. 629). Thus if the flattened cap of each sphere has a depth x, the mass of the sphere itself has decelerated over a distance $2x$. Assuming that the decelerating force is still given by the *plastic* equation, the equation of motion becomes

$$P\pi a^2 = P\pi 2rx = -m\frac{d^2}{dt^2}(2x).$$

The spheres come to rest after the time

$$t_{AD} = \frac{\pi}{2} \sqrt{\left(\frac{m}{\pi Pr}\right)}. \tag{16}$$

This is the relation given by Andrews (1930), and it is seen that the value of t_{AD} is $\sqrt{2}$ times the value given in equation (15 a). This derivation is not really valid since the elastic compression of the sphere is equal to the flattening of the cap only when the flattening is itself elastic. Under such conditions the force on the spheres cannot, of course, be expressed in terms of the plastic yield pressure P. The more completely the collision is plastic the smaller will be the overall elastic compression of the sphere *compared* with the plastic flattening at the regions of contact, and the more nearly the time of approach will be given by equation (15 a). As we are interested in the order of magnitude of the time of collision rather than in the absolute values, we may use a value between that given by equation (15 a) and that given by equation (16), say

$$t_{AD} = 1 \cdot 3 \sqrt{\left(\frac{m}{\pi Pr}\right)}. \tag{17}$$

Applying this to the experiments described by Figs. 61 b and 61 c we may calculate t_{AD}, the values of P being determined from equation (10). The results are given in Table XXI, and it is seen that there is fairly good agreement with the experimentally observed values.

TABLE XXI

Duration of Plastic Collision

Collision	Time of impact from A to D (microseconds)	
	Calculated	Observed
Steel on steel spheres .	100	150
Lead on lead spheres . .	400	600

Numerous collision experiments show that the asymmetric type of conductance curve is of general occurrence, indicating that plastic flow has occurred. Experiments also show that in the presence of thin liquid films on the surfaces, plastic flow may occur through the liquid film, without any metallic contact occurring (Rabinowicz (Thesis, 1950); Tabor, 1949). This is because the hydrodynamic pressure developed in the liquid film may easily exceed the yield pressure of the metal (Eirich and Tabor, 1948). But the time of collision is so short that there is not sufficient time for the liquid film to be squeezed out from between the surfaces.

Comparison of static and dynamic hardness

It is interesting to compare the static hardness and the dynamic hardness of metals at various velocities of impact. We may, for example, consider some impact experiments which were carried out on massive anvils of various metals using hard steel balls as the indenters (Tabor, 1948). By measuring h_1, h_2, and the chordal diameter d of the indentation formed after impact, the dynamic yield pressure P_d was calculated using equation (10). Some static experiments were also carried out and determinations made of the static yield pressure P_s required to produce impressions of the same diameter as in the corresponding impact experiments. The results showed two main points. Firstly, the dynamic yield pressure P_d was always greater than the static yield pressure P_s and the effect becomes more pronounced if we use equation (10 a) to calculate P_d instead of equation (10). This difference between P_d and P_s was particularly marked with soft metals such as lead and indium. Secondly, the dynamic yield pressure is higher at greater velocities of impact. This suggests that in calculating the dynamic yield pressure from equations (10) or (10 a), i.e. from the energy required to produce an indentation of given volume, part of the energy is used in the viscous displacement of the metal around the indentation. This view is confirmed by a calculation of the yield pressure from the height of rebound h_2. We do this using equation (6 b). This may be written in the form

$$P_r^2 = \frac{mgh_2}{2 \cdot 7 a^3}\left(\frac{1}{1/E_1 + 1/E_2}\right), \qquad (6c)$$

where the suffix r is added to P to show that it is calculated from the rebound height.

The results, which are given in Table XXII, show, at once, that although P_r is larger than P_s it is very much closer to the static value than is P_d.

TABLE XXII
Diameter of Ball 0·5 cm. Height of Fall 300 cm.

Metal	P_d/P_s	P_r/P_s
Steel	1·28	1·09
Brass	1·32	1·10
Al-alloy	1·36	1·10
Lead	1·58	1·11
Indium	5·0	1·6

The meaning of dynamic hardness

The dynamic hardness of a metal is the pressure with which it resists local indentation by a rapidly moving indenter. Under usual experimental conditions, where the speed of impact is not too large, the dynamic yield pressure is of the same order of magnitude as the static yield pressure so that, as with static hardness, dynamic hardness is essentially a measure of the elastic limit or yield stress of the metal. The actual value of the dynamic yield pressure, however, depends not only on the size of the indentation as in the Brinell test; it also depends on the velocity of impact and on the way in which it is computed. If the dynamic yield pressure (P_d) is calculated from the energy required to produce an indentation of given volume, it is larger than the static yield pressure (P_s) and increases with the velocity of impact. This is particularly marked with soft metals such as lead and indium, and it would seem that in the deformation of soft metals, where relatively large volumes of metal are displaced, appreciable forces are called into play as a result of the 'viscous' flow of the deformed metal around the indentation.

If the dynamic hardness is calculated from the height of rebound the effects of 'viscous' flow are largely eliminated. At the end of the impact process all plastic flow of the material has ended and there is no further bulk displacement of the metal around the indentation. All the deformation around the indenter is now essentially elastic and any kinetic energy imparted to the indentation is predominantly reversible. Consequently the pressures involved in this portion of the collision process (P_r) are only a few per cent. higher than those involved in the formation of indentations of the same size under static conditions.

This conclusion also shows that the large values of P_d cannot be due to work-hardening which may occur rapidly during the formation of the indentation. For at the end of the impact where the work-hardening would be a maximum, the effective yield pressure P_r is very much smaller than the mean dynamic yield pressure P_d which is involved during the course of the impact itself. This supports the view that, in the determination of P_d, forces of a quasi-viscous nature are involved.

Thus dynamic hardness values determined from rebound measurements will yield values which are close to those obtained in static measurements. Values determined from the ratio of the energy to the volume of the indentation will also be of the same order as the static values, but they will invariably be higher. With hard metals the difference will be of the order of a few per cent., but with soft metals the difference may be very much more marked and will increase with the velocity of impact. The increase in hardness or yield pressure at extremely high speeds of deformation has recently been discussed by Sir Geoffrey Taylor (1946).

Finally we may note one important implication of this work as applied to the *static* determination of hardness. The experiments described in this chapter show that extremely light impacts may be sufficient to produce plastic deformation of metals. If, therefore, in static hardness experiments, there is any jolt or impact while the load is being applied the indentation will be larger than it should be and the deduced static hardness will be lower than the true value. It is thus evident that for satisfactory static hardness measurements the load must be applied very slowly and smoothly.

REFERENCES

ANDREWS, J. P. (1930), *Phil. Mag.* **9** (7th series), 593.

—— (1931), *Proc. Phys. Soc.* **43**, 8.

BOWDEN, F. P., and TABOR, D. (1941), *Engineer*, **172**, 380.

DAVIES, R. M. (1949), *Proc. Roy. Soc.* A **197**, 416.

EDWARDS, C. A., and AUSTIN, C. R. (1923), *J. Iron & Steel Inst.* **107**, 324.

EIRICH, F. W., and TABOR, D. (1948), *Proc. Camb. Phil. Soc.* **44**, 566.

HERTZ, H. (1881), *J. reine angew. Math.* **92**, 156; see also *Miscellaneous Papers* (1896), London.

MARTEL, R. (1895), *Commission des Méthodes d'Essai des Matériaux de Construction*, Paris, **3**, 261.

OKUBO, J. (1922), *Sci. Rep. Tôhoku Univ.* **11**, 445.

PRESCOTT, J. (1927), *Applied Elasticity*, Longmans, London.

RABINOWICZ, E. (1950), Ph.D. Dissertation, Cambridge.

RAMAN, C. V. (1918), *Phys. Review*, **12**, 442; (1920), ibid. **15**, 277.

ROUDIÉ, P. (1930), *Le Contrôle de la Dureté des Métaux dans l'Industrie*, Dunod, Paris.

SHORE, A. F. (1918), *J. Iron & Steel Inst.* **2**, 59.

DE ST.-VENANT, B. (1867), *J. de Math.*, Liouville, Paris. Series 2, Vol. **12**. See also A. E. H. Love (1934), *Treatise on the Mathematical Theory of Elasticity*, Article 284, Cambridge.

TABOR, D. (1948), *Proc. Roy. Soc.* A **192**, 247; (1948), *Engineering*, **165**, 289.

—— (1949), ibid. **167**, 145.

TAGG, G. F. (1947), *J. Sci. Instr.* **24**, 244.

TAYLOR, G. I. (1946), *J. Instn. Civil Engrs.* **26**, 486 (James Forrest Lecture).

VINCENT, J. H. (1900), *Proc. Camb. Phil. Soc.* **10**, 332.

CHAPTER IX

THE AREA OF CONTACT BETWEEN SOLIDS

Asperities of hemispherical shape

WE may now discuss a more general problem which frequently
occurs in one form or another in engineering practice. When
two metal surfaces are placed in contact, what is the real area
of contact between them ? However carefully the surfaces are
prepared they will be covered with hills and valleys which are
large compared with atomic dimensions, so that the surfaces
when first placed in contact will be supported on the tips of their
asperities. The question is thus modified to a discussion of the
way in which the surface asperities are deformed when the surfaces are placed in contact. We may first assume that one
surface is harder than the other and that the asperities on the
harder surface are hemispherical at their tips. Suppose such
an asperity presses on to a flat asperity of the softer surface.
We may at once use the results discussed in Chapter IV. At
first the deformation is elastic, but for asperities of the order of
10^{-4} cm. radius the minutest loads will produce plastic deformation. Indeed full plasticity may occur even for the hardest steels
at a load of the order of a few milligrams. Consequently for
loads used in practice each asperity produces plastic flow and
behaves like a minute Brinell indenter. For a material which
is fully work-hardened the yield pressure P at the tip of each
asperity will be constant and will be given by $P \approx 3Y$, where
Y is the elastic limit of the metal. Consequently if the load
supported by the asperity is W_i, the area of contact A_i at the
tip of the asperity is given by

$$A_i = W_i/P. \tag{1}$$

The total area of contact A will be the sum of all the areas of
contact at all the asperities, so that

$$A = \sum A_i = \sum \frac{W_i}{P} = \frac{W}{P}, \tag{2}$$

where W is the total load. Thus the area of real contact will be proportional to the load, inversely proportional to the hardness of the softer metal, and *independent of the apparent area of the surfaces*. If, of course, full plasticity has not been reached at all the asperities, P will have a value less than $3Y$, but the general form of equation (2) will not be substantially altered.

If, however, the softer metal is fully annealed or only partially work-hardened, there is some change in the relation between A and W. As we saw in Chapter V, the general relation between the load W_i and the diameter of the impression d_i formed by an indenter of diameter D_i is

$$W_i = k\, \frac{d_i^n}{D_i^{n-2}},$$

so that the area of contact of the ith asperity is given by the relation

$$A_i = k'\, W_i^{2/n} D_i^{\{2(n-2)\}/n}. \tag{3}$$

If, of course, $n = 2$, as occurs for work-hardened materials, this reduces to equation (1). For any other value of n the value of A_i depends on D_i as well as on W_i. We may, however, assume that the asperities are all of the same size and shape and share the load uniformly amongst themselves, so that the total area of contact becomes

$$A = \sum A_i = \sum k'' W_i^{2/n} = k'' N W_i^{2/n} = k'' N^{(n-2)/n}(N W_i)^{2/n}$$
$$= k'' N^{(n-2)/n} W^{2/n}, \tag{3\,a}$$

where W is the total load and N is the number of points of contact. If N remains constant the relation becomes

$$A = c W^{2/n}. \tag{4}$$

Thus for fully annealed metals where n has an upper value of about $2 \cdot 5$ the relation is $A = c W^{4/5}$, and if the number N increases with the load (see later) this will tend to increase the value of A. We see, therefore, that for fully annealed metals the area of real contact A is not quite proportional to the load on account of the work-hardening accompanying indentation. If the material is partially work-hardened, as is usually the case, the work-hardening proceeds less rapidly. For example, with extruded brass or with mild steel the value of n is about $2 \cdot 15$. Consequently the real area of contact will be proportional to $W^{0.93}$ and the effect of an increase in N with W will bring this nearer to the direct proportionality between A and W.

If the hemispherical asperity is pressed against the flat surface of a harder metal, similar considerations apply. The initial elastic deformation is again restricted to the minutest loads and at very small loads full-scale plastic flow occurs. The plastic deformation of the tip again involves a yield pressure which is constant for a fully work-hardened material and has the value $P \approx 3Y$. Consequently for fully work-hardened materials the area of real contact at the tip of each asperity will be directly proportional to the load borne by the asperity. If the metal is annealed the relation is of the type $A = cW^{2/n}$, but even for fully annealed materials the index will not be less than about 0.8. Similar relations also hold for hemispherical asperities pressing on one another. For work-hardened materials the area of contact is directly proportional to W, whilst for annealed materials the area increases somewhat more slowly with W (O'Neill, 1934).

Asperities of conical and pyramidal shape

We may now consider the behaviour of asperities which are conical or pyramidal in shape. Suppose the asperities are the harder material and penetrate the plane surface of a softer metal. We may at once apply the results of Chapter VII. The tip of the asperity may be treated as a portion of a sphere of vanishingly small radius of curvature so that the elastic limit is exceeded for an infinitesimal load. Plastic deformation ensues, and for a material which does not work-harden the yield pressure P is given by $P = cY$, where c is a constant for any given indenter but varies with its angle. As we saw in Chapter VII, the more pointed the indenter the larger the value of c and the results for a steel cone penetrating work-hardened copper are redrawn in Fig. 62 (full line). It is seen that for semi-angles α ranging from 60° to 90°, c decreases from 3.6 to about 2.9. Thus for indenters of large semi-angle the constant of proportionality between P and Y does not vary rapidly with the angle and has a value of about 3.

Similar considerations apply if the pyramid or cone is of a softer metal and is squeezed flat by a harder flat surface. Here, however, the yield pressure will decrease with the semi-angle α, since a semi-angle of $0°$ is equivalent to a cylinder for which $P = Y$. Thus, as α decreases from $90°$ to $0°$, P will decrease from $2\cdot9Y$ to Y. A typical result for work-hardened copper cones

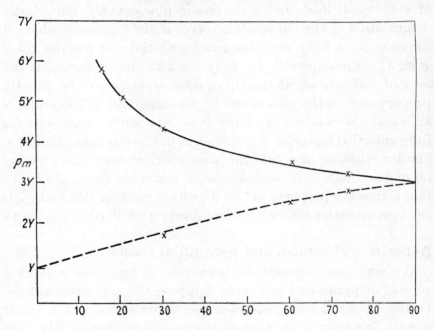

Fig. 62. Yield pressure P as a function of semi-angle α of cone. Full line—deformation of a flat surface by a hard cone. (See Fig. 46.) Broken line—deformation of a cone by a hard flat surface.

is shown dotted in Fig. 62 and confirms this conclusion. It is seen that for semi-angles α ranging from $60°$ to $90°$, the constant of proportionality between P and Y does not vary rapidly with α and has an average value of about $2\cdot7$. It is evident, therefore, that for 'blunt' asperities, that is asperities of large semi-angle, the yield pressure of a hard cone penetrating a softer metal, or of a soft cone being deformed by a harder metal, will be of the order of $3Y$.

If the deformed metal is capable of work-hardening, the amount of work-hardening will depend on the shape of the cone.

In general it will be greater for pointed than for blunt cones. As, however, the deformation produced is geometrically similar whatever its size, the yield pressure P for a cone of given angle will be constant whatever the degree of work-hardening of the metal. For cones of large semi-angle the value of P will again be approximately $3Y$, where Y is now some 'representative' measure of the elastic limit of the deformed material. Most asperities on real surfaces, in so far as they may be considered conical, are of large semi-angle. For example, even the large artificial roughnesses described in Fig. 32 (p. 64) have a semi-angle of about 60°. (It should be remembered that in this figure the vertical magnification is ten times the horizontal.)

The area of real contact

It follows that when metal surfaces are placed in contact, the tips of the asperities readily deform plastically, the mean yield pressure being given by $P = cY$. Y is some 'representative' measure of the elastic limit of the deformed metal at the tip of the asperities. The factor c depends on the shape and size of the surface irregularities, but for conical and pyramidal asperities of a wide range of angles, and for hemispherical asperities, c has a value of about 3. Consequently the yield pressure of the surface irregularities is approximately equal to $3Y$.

With conical and pyramidal asperities, the yield pressure is independent of the amount of deformation that occurs so that the real area of contact is directly proportional to the load W. This also applies to hemispherical asperities when the metals are highly worked. For annealed materials the area increases somewhat more slowly, but the difference is not large. In addition the asperities themselves are usually work-hardened by the very process of preparing the surfaces. Consequently we may expect that in most practical cases, *for a wide variety of shapes and types of surface irregularities the real area of contact A will be very nearly proportional to the applied load W*. It will also be inversely proportional to the mean yield pressure, or effective hardness P of the surface asperities. Thirdly, the real area of contact A will bear *no direct relation to the actual size of the surfaces*. In general, then, we may write

$$A = W/P. \tag{5}$$

Area of real and apparent contact

If the geometry of the surfaces is such that they make contact over a small region, and if the deformation of the surfaces is large compared with the size of the surface irregularities, the area of real contact may be essentially the same as the apparent or geometrical area of contact. This occurs, in idealized cases, in the Brinell and the Vickers indentation tests. Even here, however, the surface irregularities may persist during the deformation process and there may still be an appreciable difference between the real and apparent areas of contact. Thus, in the example quoted in Fig. 32 (p. 64), when a hard polished cylindrical indenter was pressed in to the surface of a finely grooved copper specimen, the area of real contact was about one-half the area of the macroscopic indentation produced.

This discrepancy becomes more marked when flat surfaces (or spherical surfaces of opposite curvature) are placed in contact. The apparent area of contact is the area of the surfaces themselves. The real area of contact is the summed area of all the surface irregularities which are touching and which support the load. Suppose, for example, steel flats of area 20 sq. cm. are placed in contact. The apparent area of contact will be 20 sq. cm. and it will be independent of the load. In fact, however, the surfaces will be supported on their irregularities and these will crush down until their cross-sectional area is large enough to support the load. For a steel for which P of the asperities is, say, 100 kg./mm.^{2} the area over which the asperities flow plastically will be proportional to the load and will be equal to 10 sq. mm. when the load is 1,000 kg. Thus when the surfaces are pressed together with a force of a ton the area of real contact will be 1/200th of the apparent area. For a load of 2 kg. the area of intimate contact will be 1/100,000th of the apparent area. The plastic flow of the asperities provides the real area of contact which supports the load. The stresses in the asperities are taken up by the elastic deformation of the underlying metal.

The effect of removing the load

So far we have considered the effect of applying the load. We may now consider what happens when the load is removed. It is, of course, clear that within the range of elastic deformation the process of deformation is reversible and a removal of the load enables the surfaces to return to their original configuration.

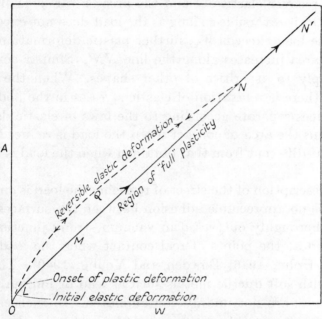

FIG. 63. Deformation of a flat surface by a hard sphere (or of a sphere by a hard flat surface). Area of contact A as a function of the load W for increasing and decreasing load. Once full plasticity has been reached (M) the area A is proportional to W (full line MNN'). For decreasing load the area is determined by the elastic equations and A is proportional to $W^{\frac{2}{3}}$ (broken line NQO).

If we consider the behaviour of a spherical asperity penetrating the surface of a softer metal we may apply the results of Chapter VI. Even if plastic flow occurs during the application of the load, the surfaces separate according to the elastic equation when the load is reduced, so that the area of contact A varies as $W^{\frac{2}{3}}$. The variation of A with W for a single asperity as the load is increased and then reduced is shown graphically in Fig. 63 for a metal which does not work-harden. Over the range OL the deformation is elastic and A is proportional to $W^{\frac{2}{3}}$.

At L the onset of plastic deformation occurs. As we saw in Chapter IV, the load at this stage for an asperity of radius of curvature of 10^{-4} cm. is less than 10^{-3} g. even for a hard steel surface. At M full plasticity is reached. The yield pressure is now almost constant and the area A increases linearly with W along the straight line MN. If the load is removed at N the surfaces recover elastically and the area A varies as $W^{\frac{2}{3}}$. The curve NQO is reversible so long as the load does not exceed W_N. When the load exceeds W_N, further plastic deformation occurs and the area increases along the line NN'. Similar considerations apply to asperities of other shapes. When the load is reduced there is relaxation of elastic stresses in the bodies and the surfaces separate according to the laws of elastic deformation. Thus the area of contact when the load is reduced will be somewhat different from that obtained when the load is initially applied.

This description of the effect of reducing the load is only valid if there is no appreciable adhesion between the surfaces. With metals thoroughly outgassed in vacuum strong junctions may be formed at the points of real contact when the load is first applied (Holm, 1946; Bowden and Young, 1949). The same occurs with soft ductile metals such as lead or indium in air if the surfaces are not unduly contaminated. In such cases the metallic junctions so formed may not be broken by the released elastic stresses when the load is reduced. Under these conditions the area of contact may not decrease appreciably when the load is removed (McFarlane and Tabor, 1950). This occurs, for example, when surfaces are joined together by pressure welding. The area of contact remains essentially the same as when the initial pressure is applied. With harder metals or with contaminated surfaces, however, the junctions appear to break readily on reducing the load and the area of contact decreases with the load according to the elastic equations.

The measurement of the area of contact

It is possible to derive a more direct measure of the area of contact between surfaces by measuring the electrical resistance between them (Holm, 1946; Bowden and Tabor, 1939). If the surfaces are free of grosser contamination the electrical resistance depends only on the electrical conductivity of the metals and on the size of the regions of contact. Suppose the metals make contact over a single circular region of radius a, and current passes from one metal to the other. The current is constricted into the region of contact, and as a result of this constriction a spreading resistance R (*Ausbreitungswiderstand*) is produced which has a value

$$R = \frac{1}{4a\lambda_1} + \frac{1}{4a\lambda_2},$$ (6)

where λ_1 and λ_2 are the specific conductivities of the two metals. In order to obtain reliable values of R it is necessary to take suitable precautions. A convenient technique is to subject the surfaces to gentle vibration (Meyer, 1898). For heavy loads this is not effective, but a slight relative movement of the surfaces produces the same effect. It is probable that this process breaks through the contaminating films at the regions of contact. By these means satisfactory and reproducible values of R may be obtained from which a may be calculated.

If the geometry is such that the region of contact is a circle and is localized in one region, the area of contact should be approximately equal to the geometric area of contact. This occurs when a spherical indenter is pressed on to a smooth flat surface, or when two cylindrical surfaces are pressed together with their axes at right angles. Under these conditions it should be possible to estimate the area of contact in three ways: from the measurements of R, from visual measurements of the permanent impressions left in the surfaces, and from the yield pressure P assuming that $A = W/P$. Some typical results for silver and steel surfaces (Bowden and Tabor, 1939) are given in Table XXIII.

Three conclusions follow from the results given in this table. First, the contact resistance is extremely small even for very light loads. For this reason considerable care must be taken in the electrical measurements. Secondly, there is good agreement

TABLE XXIII

Surfaces	Load (kg.)	R (ohms)	Area of contact (A sq. cm.)		
			From R	Visually	From P
Silver crossed cylinders	0·5	100×10^{-6}	0·0002	..	0·0002
	5·0	30×10^{-6}	0·002	..	0·002
	50·0	9×10^{-6}	0·018	0·019	0·02
	500·0	$1·9 \times 10^{-6}$	0·15	0·19	0·2
Steel crossed cylinders	1	$1·0 \times 10^{-3}$	0·00012	..	0·0001
	5	$4·9 \times 10^{-4}$	0·00061	..	0·0005
	50	$1·6 \times 10^{-4}$	0·0045	0·0045	0·005
	500	$4·9 \times 10^{-5}$	0·042	0·045	0·05
Sphere on flat (steel)	5	$4·5 \times 10^{-3}$	0·00065	..	0·0005
	50	$1·6 \times 10^{-3}$	0·0045	..	0·005
	500	$4·7 \times 10^{-5}$	0·045	..	0·05

between the three methods of determining A. This means that when the contact occurs over a single well-defined region the electrical measurements provide a reasonable measure of A, provided the surfaces have been subjected to vibration in the appropriate way. Thirdly, in such cases the area of contact is not very different for surfaces of different shapes; A is determined essentially by the load and the yield pressure.

The contact of flat surfaces

It is interesting to extend the electrical resistance measurements to the case of flat surfaces placed in contact. Some experiments of this type were carried out on steel surfaces which were carefully lapped (Bowden and Tabor, 1939). Comparison by interference fringes with optical flats showed that they were flat to within a few wave-lengths of light. One pair of flats had an area of 0·8 sq. cm., the other of 21 sq. cm. The measurements showed two striking results: (i) at any given load, the contact resistance of both pairs of flats was almost the same although the ratio of their apparent areas of contact was about 25:1; (ii) at any given load the contact resistance for the flats was of the same order of magnitude as that observed with crossed cylinders. For example, at a load of 5 kg. the contact resistance for the crossed cylinders was about 5×10^{-4} ohms, corresponding to an area of contact of about 5×10^{-4} sq. cm. For the 21 sq. cm. flat which has an apparent area 40,000 times as great, the contact resistance was only one-half this value, i.e. about $2 \cdot 5 \times 10^{-4}$ ohms. Assuming that for both types of surfaces the contact is metallic to the same degree, it is clear that only a small fraction of the area of the flat surfaces can be in intimate contact; they must touch only at the tips of the highest asperities. This would also explain the observation that the contact resistance is almost the same for the large and small flats.

It is difficult in the case of flat contacts to make an exact estimate of the real area of contact from the conductance measurements alone. The value of the conductance must depend both on the size of the metallic bridges and on their number. Since the spreading resistance of each bridge is inversely proportional to its diameter, and the area of contact is proportional to the square of the diameter, it follows that for a given resistance the area of contact is inversely proportional to the number of bridges. We do not know this number with any certainty, though in the case of flat surfaces it is clear that the number of points of contact on which they are supported cannot be less than three. If we assume the number of bridges is, in fact, three, we may calculate the area of contact from the contact resistance. The results indicate that only a very small fraction of the macroscopic area of the surfaces is in intimate contact.

It is, however, not very satisfactory to assume that the number of legs remains constant at this value. A more satisfactory type of calculation may be carried out if we assume that the yield pressure P of the asperities is approximately the same as that of the bulk metal and that the real area of contact A is determined essentially by the ratio $A = W/P$. For the steel in these experiments, $P = 100$ kg./mm.[2] If we assume that the surfaces are supported on n equal bridges of radius a, the contact resistance when the bridges are relatively far apart is given by $R = 1/2an\lambda$. Combining this equation with the relation

$$A = n\pi a^2 = W/P,$$

we obtain

$$n = \frac{\pi P}{4\lambda^2 R^2 W} = \frac{1 \cdot 39 \times 10^{-6}}{R^2 W}, \qquad (7)$$

$$a = \frac{2\lambda W R}{\pi P} = 4 \cdot 77 W R. \qquad (8)$$

Results for the 21 sq. cm. steel flats are given in Table XXIV.

TABLE XXIV

Load (kg.)	$A = W/p$ (cm.^{2})	Fraction of macroscopic area in contact	R (10^{-5} ohms)	n	a (10^{-2} cm.)
500	0·05	$\frac{1}{400}$	0·9	35	2·1
100	0·01	$\frac{1}{2000}$	2·5	22	1·2
20	0·002	$\frac{1}{10000}$	9	9	0·9
5	0·0005	$\frac{1}{40000}$	25	5	0·6
2	0·0002	$\frac{1}{100000}$	50	3	0·5

The absolute values of a and n in this table must be viewed with some reserve since the presence of a small amount of oxide will have an appreciable effect on both these parameters. For example, if half the resistance is due to an oxide film the value of n will be increased fourfold whilst the value of a will be halved. Nevertheless it is probable that the values given in the table are of the right order of magnitude. The main point brought out is that the effect of increasing the load is to increase the number and average size of the bridges. Further, it should be noted that the number of bridges is not very large even at the heaviest loads, whilst the area of the bridges lies between 10^{-3} and 10^{-4} sq. cm. These results are in complete agreement with the discussion on p. 146 and are consistent with the view that the surfaces are held apart by small irregularities which flow under the applied load until their total cross-section is sufficient to support the load. Thus the area of real contact is determined primarily by the applied load W and by the yield pressure or hardness P of the surfaces. It does not depend appreciably on the apparent size of the surfaces.

It may be worth pointing out one general uncertainty in contact resistance measurements, apart from the complication introduced by oxide or other surface films. Because of the way in which the spreading resistance arises, it cannot resolve bridges or junctions which are very close together. For example, n similar junctions (each of spreading resistance R) which are very far apart will have a spreading resistance of R/n. If they are very close together the spreading resistance will be appreciably larger (of the order R/n) and will, in fact, be very close to the spreading resistance of a single bridge embracing all the n microjunctions (Holm, 1946). Calculations based on the spreading resistance will then suggest an area of contact equal to that of the macro-bridge, although real contact may occur only over part, say one-half or two-thirds, of this area. (For a reasonable number of microjunctions the real area of contact could not be less than this without producing an appreciable change in the spreading resistance.) This would reconcile the results of Table XXIII with the talysurf records given in Fig. 32, page 64. The results in Table XXIII show good agreement between the resistance measurements and the observed indentations produced. Fig. 32, however, shows that intimate contact has occurred only over half or two-thirds of the area of the indentation. The spreading resistance of these microjunctions would differ from the spreading resistance of a single all-embracing bridge by a factor smaller than the scatter in the resistance measurements themselves. Similarly, in Table XXIV, the values of n refer to the number of macro-bridges, each of which may contain a number of microjunctions which are very close together. The values of a in the last column of this Table refer to the average over-all radius of each of these macro-bridges. It is evident that calculations of the area of contact based on resistance measurements have certain limitations but in general yield results which are of the right order of magnitude.

A fuller account of the nature of contact between surfaces is given in *Friction and Lubrication of Solids* (1950) and in Dr. Holm's interesting book, *Electrical Contacts* (1946), which provides a valuable survey of his pioneer work in this field.

REFERENCES

BOWDEN, F. P., and TABOR, D. (1939), *Proc. Roy. Soc.* A **169**, 391.

────── ────── (1950), *Friction and Lubrication of Solids*, Clarendon Press.

────── and YOUNG, J. E. (1949), *Nature*, **164**, 1089.

HOLM, R. (1946), *Electrical Contacts*, Almquist and Wiksells, Stockholm.

MCFARLANE, J. W., and TABOR, D. (1950), in press.

MEYER, A. (1898), *Öfvers. Vetensk. Akad. Förh., Stockh.* **55**, 199.

O'NEILL, H. (1934), *The Hardness of Metals and Its Measurements*, Chapman and Hall, London.

APPENDIX

I. BRINELL HARDNESS

Definition

A HARD spherical indenter of diameter D mm. is pressed into the metal surface under a load W kg. and the mean chordal diameter of the resultant indentation measured (d mm.). The Brinell hardness number (B.H.N.) is defined as

$$\text{B.H.N.} = \frac{W}{\text{curved area of indentation}}$$

$$= \frac{2W}{\pi D\{D - \sqrt{(D^2 - d^2)}\}}$$

and is expressed in kg./mm.²

Size of indentation

For most metals the B.H.N. depends on the size of the indentation. However, geometrically similar indentations (i.e. those for which the ratio d/D is constant) give the same B.H.N. To a first approximation this occurs when the ratio W/D^2 is kept constant. It is usual to use indentation sizes ranging from about $d/D = 0.3$ to $d/D = 0.6$.

Specified loads

If a single hardness value is required for specification purposes the British Standards Association specifies the following loads for various metals (Table Ia).

TABLE I (a)

		Load W kg. for			
	W/D^2	$D = 1$ mm.	$D = 2$ mm.	$D = 5$ mm.	$D = 10$ mm.
Iron, steel, materials of similar hardness	30	30	120	750	3,000
Brasses, bronzes . . .	10	10	40	250	1,000
Pure copper	5	5	20	125	500
Lead-, tin-alloys . . .	1	1	4	25	100

Loads for validity of Meyer's law

Meyer's law states that $W = kd^n$, where k is a constant and n another constant for the metal. The value of n depends on the degree of work-hardening of the metal; for annealed metals it is near 2.5, for work-hardened metals near 2. If the loads are too small the deformation is not fully plastic and deviations from this relation are observed, the value of n being larger than its true value. For Meyer's relation to be valid with a ball of 10 mm. diameter the loads should exceed the values given in Table Ib. (For fuller details see Chapter IV, pp. 51–4.)

TABLE I (*b*)

Approx. B.H.N. of metal (kg./mm.²)		For Meyer's law to be valid with 10-mm. ball, load should exceed
Lead-, tin-	10	1 kg.
alloys	50	30 ,,
Copper alloys,	100	50 ,,
	200	180 ,,
Steels	400	1,500 ,,
	600	5,200 ,,

With balls of different diameters the loads will be proportional to the square of the diameter. Thus for a 5-mm. ball the loads for Meyer's law to be valid will be about one quarter of those given in Table I *b*, whilst for a ball of 1 mm. diameter the values will be about one hundredth.

Time of loading

The time of loading is usually 30 sec. for steels and 60 sec. for brasses and bronzes.

Specimen size

The specimen should be large enough to ensure that all the plastic flow occurs within a region considerably smaller than the specimen itself. A working standard is to use a thickness of specimen at least 10 times the depth of the indentation for hard metals and at least 15 times the depth for softer metals. The specimen should also be about 4 times wider than the diameter of the impression. In general, if a ball of 10 mm. diameter is used, the specimen should not be less than about $\frac{1}{2}$ in. wide and $\frac{1}{2}$ in. deep.

TABLE I (c) *Brinell Hardness Numbers for 10-mm. Ball, 3,000 kg. and 500 kg. Load*

(For a given size of indentation the B.H.N. is directly proportional to the load W. For this reason, only two loads are given in this table.)

Diam. of indentation (mm.)	Curved area of indentation (mm.²)	B.H.N. (kg./mm.²) W kg. 3,000	B.H.N. (kg./mm.²) W kg. 500	Diam. of indentation (mm.)	Curved area of indentation (mm.²)	B.H.N. (kg./mm.²) W kg. 3,000	B.H.N. (kg./mm.²) W kg. 500
1·50	1·777	1,688	281	4·25	14·893	201	34
·55	1·898	1,580	263	·30	15·264	197	33
·60	2·024	1,482	247	·35	15·640	192	32
·65	2·153	1,393	232	·40	16·022	187	31
·70	2·287	1,312	219	·45	16·410	183	30
·75	2·424	1,238	206	·50	16·803	179	30
·80	2·566	1,169	195	·55	17·202	174	29
·85	2·711	1,107	184	·60	17·606	170	28
·90	2·861	1,049	175	·65	18·016	167	28
·95	3·014	995	166	·70	18·432	163	27
				·75	18·853	159	27
2·00	3·174	946	158	·80	19·280	156	26
·05	3·335	900	150	·85	19·712	152	25
·10	3·502	857	143	·90	20·150	149	25
·15	3·673	817	136	·95	20·594	146	24
·20	3·848	780	130				
·25	4·027	745	124	5·00	21·044	143	24
·30	4·211	712	119	·05	21·501	140	23
·35	4·399	682	114	·10	21·964	137	23
·40	4·591	653	109	·15	22·433	134	22
·45	4·787	627	104	·20	22·907	131	22
·50	4·988	601	100	·25	23·389	128	21
·55	5·192	578	96·5	·30	23·877	126	21
·60	5·401	555	93	·35	24·370	123	21
·65	5·615	534	89	·40	24·870	121	20
·70	5·834	514	86	·45	25·377	118	20
·75	6·055	495	83	·50	25·891	116	19
·80	6·283	477	80	·55	26·413	114	19
·85	6·514	461	77	·60	26·940	111	19
·90	6·750	444	74	·65	27·475	109	18
·95	6·990	429	72	·70	28·016	107	18
				·75	28·564	105	18
3·00	7·235	415	69	·80	29·119	103	17
·05	7·485	401	67	·85	29·682	101	17
·10	7·737	388	65	·90	30·254	99	17
·15	7·997	375	63	·95	30·831	97	16
·20	8·260	363	61				
·25	8·526	352	59	6·00	31·416	95	16
·30	8·800	341	57	·05	32·009	94	16
·35	9·076	331	55	·10	32·610	92	15
·40	9·358	321	53	·15	33·219	90	15
·45	9·644	311	52	·20	33·836	89	15
·50	9·934	302	50	·25	34·459	87	15
·55	10·230	293	49	·30	35·092	85	14
·60	10·532	285	47	·35	35·734	84	14
·65	10·838	277	46	·40	36·384	82	14
·70	11·148	269	45	·45	37·042	81	13
·75	11·464	262	44	·50	37·708	80	13
·80	11·783	255	42	·55	38·386	78	13
·85	12·108	248	41	·60	39·072	77	13
·90	12·438	241	40	·65	39·766	75	13
·95	12·774	235	39	·70	40·469	74	12
				·75	41·183	73	12
4·00	13·113	229	38	·80	41·907	72	12
·05	13·460	223	37	·85	42·640	70	12
·10	13·808	217	36	·90	43·384	69	12
·15	14·164	212	35	·95	44·139	68	11
·20	14·526	207	34	7·00	44·903	67	11

The curved areas of the indentations given in Wahlberg's paper (1901) are not reliable and have been very kindly recalculated for these tables by Dr. E. Rabinowicz.

II. MEYER HARDNESS

Definition

A HARD spherical indenter of diameter D mm. is pressed into the surface of the metal specimen under a load W kg. and the mean chordal diameter of the resultant indentation measured (d mm.). The Meyer hardness number (M.H.N.) is defined as

$$\text{M.H.N.} = \frac{W}{\text{projected area of indentation}}$$

$$= \frac{4W}{\pi d^2}$$

and is expressed in kg./sq. mm. The practical details of measurement are the same as for Brinell hardness measurements.

Meyer hardness and Brinell hardness

For a given indentation obtained under a given load, the Meyer hardness is always greater than the corresponding Brinell hardness by the ratio of the curved to the projected area of the indentation. The value of this ratio is given in Table II a.

TABLE II (a)

Size of indentation d/D	Ratio: M.H.N./B.H.N.
0·1	1·002
0·15	1·006
0·20	1·010
0·25	1·016
0·30	1·024
0·35	1·032
0·40	1·044
0·45	1·057
0·50	1·074
0·55	1·090
0·60	1·111
0·65	1·136
0·70	1·167
0·75	1·204
0·80	1·250
0·85	1·310
0·90	1·393
0·95	1·524
1·00	2·000

For indentations smaller than $d/D = 0.4$ the following approximate relation may be used:

$$\text{Ratio: M.H.N./B.H.N.} = 1 + \tfrac{1}{4}(d/D)^2.$$

TABLE II (b)

Meyer Hardness Numbers for 10-mm. Ball, 3,000 kg. and 500 kg. Load

(The M.H.N. is directly proportional to the load. Values are therefore only given for two loads.)

Diam. of indentation (mm.)	Projected area of indentation (mm.^{2})	M.H.N. (kg./mm.^{2})		Diam. of indentation (mm.)	Projected area of indentation (mm.^{2})	M.H.N. (kg./mm.^{2})	
		3,000 kg.	500 kg.			3,000 kg.	500 kg.
2·00	3·1416	955	159	4·55	16·260	185	31
·05	3·3006	909	152	·60	16·619	181	30
·10	3·4636	867	144	·65	16·982	177	29
·15	3·6305	827	138	·70	17·349	173	29
·20	3·8013	790	132	·75	17·720	169	28
·25	3·9761	754	122	·80	18·096	166	28
·30	4·1548	723	121	·85	18·474	162	27
·35	4·3374	692	115	·90	18·857	159	27
·40	4·5239	664	111	·95	19·244	156	26
·45	4·7143	637	106				
·50	4·9087	613	102	5·00	19·635	153	25
·55	5·1070	588	98	·05	20·003	150	25
·60	5·3093	566	94	·10	20·428	147	24
·65	5·5155	544	91	·15	20·831	144	24
·70	5·7256	524	87	·20	21·237	141	24
·75	5·9396	506	84	·25	21·647	139	23
·80	6·1575	487	81	·30	22·062	136	23
·85	6·3794	470	78	·35	22·480	133	22
·90	6·6052	454	76	·40	22·902	131	22
·95	6·8449	438	73	·45	23·328	129	21
				·50	23·758	126	21
3·00	7·0686	425	71	·55	24·192	124	21
·05	7·3062	411	68	·60	24·630	122	20
·10	7·5477	397	66	·65	25·072	120	20
·15	7·7931	385	64	·70	25·518	118	20
·20	8·0425	373	62	·75	25·967	116	19
·25	8·2958	362	60	·80	26·421	114	19
·30	8·5530	351	58	·85	26·878	112	19
·35	8·8141	341	57	·90	27·340	110	18
·40	9·0792	330	55	·95	27·805	108	18
·45	9·3482	321	54				
·50	9·6211	312	52	6·00	28·274	106	18
·55	9·8980	303	50	·05	28·747	104	17
·60	10·179	295	49	·10	29·225	103	17
·65	10·463	287	48	·15	29·706	101	17
·70	10·752	279	47	·20	30·191	99·4	17
·75	11·045	272	45	·25	30·680	97·8	16
·80	11·341	265	44	·30	31·172	96·3	16
·85	11·642	258	43	·35	31·669	94·8	16
·90	11·946	251	42	·40	32·170	93·3	16
·95	12·254	245	41	·45	32·674	91·9	15
				·50	33·183	90·4	15
4·00	12·566	239	40	·55	33·696	89·0	15
·05	12·882	233	39	·60	34·212	87·7	15
·10	13·203	227	38	·65	34·732	86·4	14
·15	13·526	222	37	·70	35·257	85·2	14
·20	13·854	216	36	·75	35·785	83·9	14
·25	14·186	211	35	·80	36·317	82·7	14
·30	14·522	207	34	·85	36·853	81·5	14
·35	14·862	202	34	·90	37·393	80·2	13
·40	15·205	197	33	·95	37·937	79·1	13
·45	15·553	193	32				
·50	15·904	189	31	7·00	38·485	78·0	13

III. VICKERS HARDNESS

Definition

A PYRAMIDAL diamond indenter is pressed into the surface of the metal under a load of W kg. and the mean diagonal of the resultant indentation measured (d mm.). The Vickers hardness number, or Vickers diamond hardness (V.D.H.), is defined as

$$\text{V.D.H.} = \frac{W}{\text{pyramidal area of indentation}}.$$

The indenter has an angle of 136° between opposite faces and 146° between opposite edges. From simple geometry this means that the pyramidal area of the indentation is greater than the projected area of the indentation by the ratio $1:0.9272$. Hence

$$\text{V.D.H.} = \frac{0.9272\,W}{\text{projected area of indentation}}$$

$$= 1.8544 W/d^2.$$

The value is expressed in kg./mm.²

Vickers hardness and mean pressure

The relation between the Vickers hardness and the mean yield pressure over the indentation (P_m) is simply

$$\text{V.D.H.} = 0.9272 P_m.$$

Loading

The loads used usually range from 1 to 120 kg. Since the indentation is geometrically similar whatever its size, the Vickers hardness number is practically independent of the load.

Time of loading

The load is generally applied for a period of 10 sec. or longer.

TABLE III (a)

Vickers Diamond (Pyramid) Hardness for 10-kg. Load

(Note that for a given size of indentation the V.D.H. is directly proportional to the load W. For this reason, only one load is given in this table. Note also that with the $\frac{2}{3}$-in. objective, each scale division equals 0·001 mm., whilst with the $1\frac{1}{2}$-in. objective, each scale division is equal to 0·0025 mm.)

Diagonal of indentation 0·001 mm.	VICKERS hardness number (kg./mm.²). Load = 10 kg.									
	0	1	2	3	4	5	6	7	8	9
100	1,855	1,818	1,783	1,749	1,715	1,682	1,650	1,619	1,589	1,561
110	1,533	1,505	1,478	1,452	1,427	1,402	1,378	1,354	1,332	1,310
120	1,288	1,267	1,246	1,226	1,206	1,187	1,168	1,150	1,132	1,115
130	1,097	1,081	1,064	1,048	1,033	1,018	1,003	988	974	960
140	946	933	920	907	894	882	870	858	847	835
150	824	813	803	792	782	772	762	752	743	734
160	724	715	707	698	690	681	673	665	657	649
170	642	634	627	620	613	606	599	592	585	579
180	572	566	560	554	548	542	536	530	525	519
190	514	508	503	498	493	488	483	478	473	468
200	464	459	455	450	446	442	437	433	429	425
210	421	417	413	409	405	401	397	394	390	387
220	383	380	376	373	370	366	363	360	357	354
230	351	348	345	342	339	336	333	330	327	325
240	322	319	317	314	312	309	306	304	302	299
250	297	294	292	289	287	285	283	281	279	276
260	274	272	270	268	266	264	262	260	258	256
270	254	253	251	249	247	245	243	242	240	238
280	236	235	233	232	230	228	227	225	224	222
290	221	219	218	216	215	213	212	210	209	207
300	206	205	203	202	201	199	198	197	196	194
310	193	192	191	189	188	187	186	185	183	182
320	181	180	179	178	177	176	175	173	172	171
330	170	169	168	167	166	165	164	163	162	161
340	160	160	159	158	157	156	155	154	153	152
350	151	151	150	149	148	147	146	146	145	144
360	143	142	142	141	140	139	138	138	137	136
370	136	135	134	133	133	132	131	131	130	129
380	128	128	127	126	126	125	124	124	123	123
390	122	121	121	120	120	119	118	118	117	117
400	116	115	115	114	114	113	113	112	111	111
410	110	110	109	109	108	108	107	107	106	106
420	105	105	104	104	103	103	102	102	101	101

TABLE III (a) *continued*

Diagonal of indentation 0·001 mm.	VICKERS *hardness number* (kg./mm.²). Load = 10 kg.									
	0	1	2	3	4	5	6	7	8	9
430	100	99·8	99·4	98·9	98·5	98·0	97·6	97·1	96·7	96·2
440	95·8	95·3	94·9	94·5	94·1	93·6	93·2	92·8	92·4	92·0
450	91·6	91·2	90·8	90·4	90·0	89·6	89·2	88·8	88·4	88·0
460	87·6	87·3	86·9	86·5	86·1	85·8	85·4	85·0	84·7	84·3
470	84·0	83·6	83·2	82·9	82·5	82·2	81·8	81·5	81·2	80·8
480	80·5	80·2	79·8	79·5	79·2	78·8	78·5	78·2	77·9	77·6
490	77·2	76·9	76·6	76·3	76·0	75·7	75·4	75·1	74·8	74·5
500	74·2	73·9	73·6	73·3	73·0	72·7	72·4	72·1	71·9	71·6
510	71·3	71·0	70·7	70·5	70·2	69·9	69·6	69·4	69·1	68·8
520	68·6	68·3	68·1	67·8	67·5	67·3	67·0	66·8	66·5	66·3
530	66·0	65·8	65·5	65·3	65·0	64·8	64·5	64·3	64·1	63·8
540	63·6	63·4	63·1	62·9	62·7	62·4	62·2	62·0	61·7	61·5
550	61·3	61·1	60·9	60·6	60·4	60·2	60·0	59·8	59·6	59·3
560	59·1	58·9	58·7	58·5	58·3	58·1	57·9	57·7	57·5	57·3
570	57·1	56·9	56·7	56·5	56·3	56·1	55·9	55·7	55·5	55·3
580	55·1	54·9	54·7	54·6	54·4	54·2	54·0	53·8	53·6	53·4
590	53·3	53·1	52·9	52·7	52·6	52·4	52·2	52·0	51·9	51·7
600	51·5	51·3	51·2	51·0	50·8	50·7	50·5	50·3	50·2	50·0
610	49·8	49·7	49·5	49·4	49·2	49·0	48·9	48·7	48·6	48·4
620	48·2	48·1	47·9	47·8	47·6	47·5	47·3	47·2	47·0	46·9
630	46·7	46·6	46·4	46·3	46·1	46·0	45·8	45·7	45·6	45·4
640	45·3	45·1	45·0	44·8	44·7	44·6	44·4	44·3	44·2	44·0
650	43·9	43·8	43·6	43·5	43·4	43·2	43·1	43·0	42·8	42·7
660	42·6	42·4	42·3	42·2	42·1	41·9	41·8	41·7	41·6	41·4
670	41·3	41·2	41·1	40·9	40·8	40·7	40·6	40·5	40·3	40·2
680	40·1	40·0	39·9	39·8	39·6	39·5	39·4	39·3	39·2	39·1
690	39·0	38·8	38·7	38·6	38·5	38·4	38·3	38·2	38·1	38·0
700	37·8	37·7	37·6	37·5	37·4	37·3	37·2	37·1	37·0	36·9
710	36·8	36·7	36·6	36·5	36·4	36·3	36·2	36·1	36·0	35·9
720	35·8	35·7	35·6	35·5	35·4	35·3	35·2	35·1	35·0	34·9
730	34·8	34·7	34·6	34·5	34·4	34·3	34·2	34·1	34·0	34·0
740	33·9	33·8	33·7	33·6	33·5	33·4	33·3	33·2	33·1	33·1
750	33·0	32·9	32·8	32·7	32·6	32·5	32·4	32·4	32·3	32·2
760	32·1	32·0	31·9	31·8	31·8	31·7	31·6	31·5	31·4	31·4
770	31·3	31·2	31·1	31·0	30·9	30·9	30·8	30·7	30·7	30·6
780	30·5	30·4	30·3	30·3	30·2	30·1	30·0	29·9	29·9	29·8
790	29·7	29·6	29·6	29·5	29·4	29·3	29·3	29·2	29·1	29·1
800	29·0	28·9	28·8	28·8	28·7	28·7	28·6	28·5	28·4	28·3
810	28·3	28·2	28·1	28·0	28·0	27·9	27·8	27·8	27·7	27·7

TABLE III (*a*) *continued*

Diagonal of indentation 0·001 mm.	VICKERS *hardness number* (kg./mm.²). *Load* = 10 kg.									
	0	1	2	3	4	5	6	7	8	9
820	27·6	27·5	27·4	27·4	27·3	27·3	27·2	27·1	27·0	27·0
830	26·9	26·8	26·8	26·7	26·7	26·6	26·5	26·5	26·4	26·3
840	26·3	26·2	26·2	26·1	26·0	26·0	25·9	25·8	25·8	25·7
850	25·7	25·6	25·6	25·5	25·4	25·4	25·3	25·3	25·2	25·1
860	25·1	25·0	25·0	24·9	24·8	24·8	24·7	24·7	24·6	24·6
870	24·5	24·4	24·4	24·3	24·3	24·2	24·2	24·1	24·1	24·0
880	24·0	23·9	23·8	23·8	23·7	23·7	23·6	23·6	23·5	23·5
890	23·4	23·4	23·3	23·3	23·2	23·2	23·1	23·0	23·0	22·9
900	22·9	22·8	22·8	22·7	22·7	22·6	22·6	22·5	22·5	22·4
910	22·4	22·3	22·3	22·3	22·2	22·2	22·1	22·1	22·0	22·0
920	21·9	21·9	21·8	21·8	21·7	21·7	21·6	21·6	21·5	21·5
930	21·4	21·4	21·4	21·3	21·3	21·2	21·2	21·1	21·1	21·0
940	21·0	20·9	20·9	20·8	20·8	20·8	20·7	20·7	20·6	20·6
950	20·5	20·5	20·5	20·4	20·4	20·3	20·3	20·2	20·2	20·2
960	20·1	20·1	20·0	20·0	20·0	19·9	19·9	19·8	19·8	19·8
970	19·7	19·7	19·6	19·6	19·6	19·5	19·5	19·4	19·4	19·4
980	19·3	19·3	19·2	19·2	19·2	19·1	19·1	19·0	19·0	19·0
990	18·9	18·9	18·8	18·8	18·8	18·7	18·7	18·7	18·6	18·6

IV. HARDNESS CONVERSION

THE values given here are approximate only and apply only to steels of uniform chemical composition and uniform heat treatment. The values are less reliable for non-ferrous materials and are not recommended for case-hardened steels. The greater part of the data is taken from the A.S.M. Handbook.

TABLE IV (a)

Hardness Conversion Table

BRINELL 10-mm. steel ball load 3,000 kg.		VICKERS diamond pyramid hardness $(kg./mm.^2)$	ROCKWELL		SHORE rebound scleroscope no.	MOHS'S scale	Approx. ultimate tensile strength $(tons/in.^2)$
Diam. (mm.)	Hardness $(kg./mm.^2)$		C 150 kg. load 120° diamond cone	B 100 kg. load $\frac{1}{16}$-in. steel ball			
2·20	780*	1,150	70	..	106	8·5	~ 200
·25	745*	1,050	68	..	100	..	..
·30	712*	960	66	..	95	..	..
·35	682*	885	64	..	91	..	..
·40	653*	820	62	..	87	8·0	..
·45	627*	765	60	..	84	..	..
·50	601*	717	58	..	81	..	~ 150
·55	578*	675	57	..	78	7·5	..
·60	555*	633	55	~ 120	75	..	..
·65	534*	598	53	..	72	..	..
·70	514*	567	52	..	70	..	..
·75	495*	540	50	..	67	..	..
·80	477*	515	49	..	65	7·0	..
·85	461*	494	47	..	63	..	102
·90	444	472	46	..	61	..	98
·95	429	454	45	~ 115	59	..	95
3·00	415	437	44	..	57	..	91
·05	401	420	42	..	55	6·5	88
·10	388	404	41	..	54	..	84
·15	375	389	40	..	52	..	81
·20	363	375	38	~ 110	51	..	79
·25	352	363	37	..	49	..	76
·30	341	350	36	..	48	..	74
·35	331	339	35	..	46	6·0	71
·40	321	327	34	..	45	..	69
·45	311	316	33	..	44	..	67
·50	302	305	32	..	43	..	65
·55	293	296	31	..	42	..	63
·60	285	287	30	105	40	..	62
·65	277	279	29	104	39	5·5	60
·70	269	270	28	104	38	..	58
·75	262	263	26	103	37	..	57
·80	255	256	25	102	37	..	56
·85	248	248	24	102	36	..	55
·90	241	241	23	100	35	..	53
·95	235	235	22	99	34	..	52
4·00	229	229	21	98	33	..	50
·05	223	223	20	97	32	..	49
·10	217	217	18	96	31	..	48
·15	212	212	17	96	31	..	47
·20	207	207	16	95	30	5·0	45

* With the standard steel ball, Brinell hardness values above about 400 are not considered reliable. See Chapter IV, especially p. 55.

TABLE IV (a) *continued*

BRINELL 10-mm. steel ball load 3,000 kg.		VICKERS diamond pyramid hardness (kg./mm.²)	ROCKWELL		SHORE rebound scleroscope no.	MOHS'S scale	Approx. ultimate tensile strength (tons/in.²)
Diam. (mm.)	Hardness (kg./mm.²)		C 150 kg. load 120° diamond cone	B 100 kg. load $\frac{1}{16}$-in. steel ball			
4·25	201	201	15	94	30	..	44
·30	197	197	13	93	29	..	43
·35	192	192	12	92	28	..	42
·40	187	187	10	91	28	..	42
·45	183	183	9	90	27	..	41
·50	179	179	8	89	27	..	40
·55	174	174	7	88	26	..	39
·60	170	170	6	87	26	..	38
·65	167	167	4	86	25	..	37
·70	163	163	3	85	25	..	37
·75	159	159	2	84	24	4·5	36
·80	156	156	1	83	24	..	35
·85	152	152	..	82	23	..	34
·90	149	149	..	81	23	..	33
·95	146	146	..	80	22	..	33
5·00	143	143	..	79	22	..	32
·05	140	140	..	78	21	..	32
·10	137	137	..	77	21	..	31
·15	134	134	..	76	21	..	30
·20	131	131	..	74	20	..	29
·25	128	128	..	73	20	..	29
·30	126	126	..	72	..	..	29
·35	123	123	..	71	..	..	28
·40	121	121	..	70	..	..	28
·45	118	118	..	69	..	..	27
·50	116	116	..	68	..	..	27
·55	114	114	..	67	..	..	26
·60	111	111	..	66	..	..	26
·65	109	109	..	65	..	..	25
·70	107	107	..	64	..	..	25
·75	105	105	..	62	..	..	24
·80	103	103	..	61	..	..	24
·85	101	101	..	60	..	..	23
·90	99	99	..	59	..	..	23
·95	97	97	..	57	..	..	22
6·00	96	96	..	56	..	..	22

V. HARDNESS AND
ULTIMATE TENSILE STRENGTH

IN Table V (a) approximate values are given for the ratio C, where

Ultimate tensile strength $= C \times$ Brinell hardness number.

Since the B.H.N. is always given in kg./mm.^{2}, the constant C has different values depending on whether the ultimate tensile strength is measured in tons/in.^{2} or in kg./mm.^{2} It should be emphasized that these values, which are taken largely from Greaves and Jones and from O'Neill, are only approximate since C depends appreciably on the degree of work-hardening of the metal. (For further details see Chapter V.)

TABLE V (a)

Hardness and Ultimate Tensile Strength

Metal	Condition	C tons/in.^{2}	C kg./mm.^{2}
Non-Ferrous			
Aluminium	Annealed	0·33–0·36	0·52–0·57
	Drawn	0·27	0·43
	Heavily worked	0·21–0·23	0·33–0·36
Brass, bronze	Annealed	0·33–0·36	0·52–0·57
	Cold-worked	0·26	0·41
Copper	Annealed	0·33–0·36	0·52–0·57
	Cold-worked	0·24–0·26	0·38–0·41
Duralumin	Annealed	0·23–0·24	0·36–0·38
	Aged	0·22–0·23	0·34–0·36
Lead	Cast, rolled	0·27	0·43
Nickel	Annealed	0·31	0·49
	Cold-worked	0·26	0·41
Tin	Cast, rolled	0·29	0·46
Ferrous			
Armco iron	Annealed	0·22	0·35
Alloy steels	Heat-treated B.H.N. < 250	0·215	0·34
	B.H.N. 250–400	0·21	0·33
Carbon steels	Rolled, normalized, annealed	0·22	0·35
(medium)	Heat-treated	0·215	0·34
Mild steel	Rolled, normalized, annealed	0·23	0·36

VI. SOME TYPICAL HARDNESS VALUES

THE hardness of metals and alloys depends on their exact composition, their crystal size, and in particular on their degree of work-hardening. For this reason the values given in Table VI (a) should be considered as typical values rather than as absolute values. Most of the data are taken from the A.S.M. Handbook (1948 edition).

TABLE VI (a)

Typical Hardness Values

Metal or alloy	Condition	B.H.N. (kg./mm.^{2})	Ultimate tensile strength (tons/in.^{2})
Aluminium 99·97	Annealed	16	3–4
	Cold rolled 75%	27–30	7–8
99·5	Cast	20	5
	Hot rolled	30	7
	Cold rolled	40	9
99·3	Cold rolled	45	10–11
Aluminium–magnesium Al 94 Mg 6	Rolled	70	18
Aluminium–manganese Al 98·8 Mn 1·2	Annealed	28	8
	Rolled	55	12–14
Aluminium–nickel Al 90 Ni 10	Rolled	50	11
Aluminium–zinc–magnesium–copper Zn 6·4 Mg 2·5 Cu 1·2 remainder Al	Annealed	60	14
	Quenched and aged	170	36
Antimony	Cast	30–60	..
Arsenic	Cast	147	..
Beryllium–copper Be 2 Cu 98	Quenched	150	33
	Quenched and rolled	220	50
	Quenched and precipitation hardened	360–400	80–90
Bismuth	Cast	11–12	..
Brass α : Cu 70 Zn 30 (cartridge)	Annealed	80–100	20
	Rolled hard	140	30
	Spring	190	42
β : Cu 60 Zn 40 (Muntz)	Annealed	80–100	25
	Hard	140	30
Bronze Cu 96 Sn 4	Cast and annealed	60	14
Cu 90 Sn 10 (*see* phosphor–bronze)	Cast and annealed	80	19
Cadmium	Cast	23	5–6
Calcium	Cast	17	4
Cerium	..	28	..

TABLE VI (*a*) *continued*

Metal or alloy	Condition	B.H.N. (kg./mm.²)	Ultimate tensile strength (tons/in.²)
Chromel			
A. Ni 82·5 Cr 15 Fe 1	Rolled hot	175–210	50–56
B. Ni 77·5 Cr 20 Fe 1	Rolled hot	180–220	50–57
C. Ni 61 Cr 12 Fe 25	Rolled hot	180–200	40–50
Chromium	Cast	100–170	..
,, electroplate	Hard bright	500–1,250	..
	Annealed	∼ 100	..
Cobalt	Annealed	48	16
,,	Drawn	140	43
,, electroplate	Bright	300	..
Constantan	Annealed	80–100	25–30
Ni 45 Cu 55	Cold rolled	120–300	30–60
Copper, pure	Annealed	30–40	9–12
	Heavily worked	100–120	22–28
,, high conductivity	Annealed	40	12–14
	Hard sheet	80	20–24
	Hard drawn wire	90–130	20–28
Copper–aluminium bronzes			
Cu 95 Al 5. α form	Hot rolled	124	26
Cu 90 Al 10. β form	Cast	130	30–32
	Hot rolled	200	40–45
Cu 82·5 Al 10 Ni 5 Fe 2·5	Heat treated	250	50
Copper–lead bearing alloys			
Pb 27, non-dendritic	Cast	35	..
Pb 20, dendritic	Cast	25	..
Copper–nickel alloys			
Cu 55 Ni 18 Zn 27	Annealed	90	25
(nickel silver)	Cold rolled	220	50
Cu 70 Ni 30	'Soft'	50–70	20–25
(cupro-nickel)	Hard worked	140	35
Duralumin	Annealed	40	8–10
Al 93·5 Cu 4·4 Mg 1·5	Heat treated	100	25
Mn 0·6	Cold rolled	120	30
Gallium, pure	Cast	6–7	..
Gold, pure	Cast	30	8–10
	Hard drawn	60–70	15–16
Gun-metal	Annealed	50–100	11–22
	Work-hardened	190	45
Indium, pure	Cast, rolled	1	..
Iridium, pure	Cast	170	..
	Hard rolled	350	..

TABLE VI (a) *continued*

Metal or alloy	Condition	B.H.N. (kg./mm.²)	Ultimate tensile strength (tons/in.²)
Iron, pure	Electrolytic, annealed	70	20
	,, worked	200	45
,, cast	Chilled	450	..
	Grey	150–240	..
	Centrifugally cast	200	..
	'Whiteheart' malleable	150–200	..
Lanthanum	..	37	..
Lead 99·9	Cast	4	0·8
	Rolled	4	1·4
Lead–antimony (hard lead) Pb 96 Sb 4	Rolled	8	2
	Quenched and aged	24	5
Lead–tin (soft solders) Pb 30 Sn 70	Cast	12	3
Pb 37 Sn 63	Cast	14	4
Lead-bearing alloy Pb 78·5, Sb 15, Sn 6, Cu 0·5	Cast	21	..
Magnesium 99·98	Sand cast	30	6·7
	Drawn and annealed	40	10–12
	Hard drawn	50	12–16
Magnesium–aluminium Mg 91·8 Al 8 Mn 0·2 (Dow metal A)	Cast	50	14
Magnesium–silicon (Si 1·2)	Drawn	80	18
Magnesium–zinc (Zn 3 Al 0·5)	Drawn	80	20
Manganese	γ form, quenched	300	30
Manganese–bronze (medium)	Worked	110–120	29–37
(high tensile)	Heavily worked	200–270	45–50
Molybdenum	Drawn	160–180	30–40
	Sintered	160	..
Monel metal Ni 67 Cu 30	Cast	100–140	32–37
	Cold drawn	160–250	37–56
	Spring temper	∼ 300	60–75
Nickel 99·95	Annealed	70–80	20
	Electroplate	200–400	..
99·4	Annealed	90–120	22–30
	Drawn	125–230	30–55
Nickel–iron (C < 0·1) Ni 36 Fe 64 (Invar)	Rolled	160	38
Ni 43 Fe 57 (magnetic)	Rolled	200	45
Nickel–manganese Ni 95 Mn 4·5	Hot rolled	150	38

TABLE VI (a) *continued*

Metal or alloy	Condition	B.H.N. $(kg./mm.^2)$	Ultimate tensile strength $(tons/in.^2)$
Osmium, pure	Cast	400	..
Palladium (99·9+)	Annealed	40	10
	Drawn	100	22
Phosphor–bronze A	Soft	90	20–25
Sn 3·8–5·8, P 0·03–0·35	Hard	150–180	30–40
remainder Cu.	Spring	190–220	44–50
Platinum, pure	Annealed	40	9
	Drawn	100	21
	Electroplated	~ 600	..
Platinum–iridium			
Pt 95 Ir 5	Annealed	90	18
	Hard worked	140	32
Pt 90 Ir 10	Annealed	130	25
	Hard worked	190	40
Pt 75 Ir 25	Annealed	240	56
	Hard worked	310	80
Platinum–rhodium	Annealed	90	20
Pt 90 Rh 10	Hard worked	170	40
Potassium	Cast	0·037	..
Praesodymium	..	25	..
Rhodium	Cast, annealed	135	35
	Hard worked	390	90
	Electroplated	800	..
Ruthenium	Cast	220	..
Silver, pure	Annealed at 650° C.	25	8–10
	Hard drawn	80	20–23
	Electroplated	~ 100	..
Silver–copper	Quenched	50	16–20
Ag 92·5 Cu 7·5	Quenched and aged	100–140	27–30
(sterling)			
Sodium	Cast	0·07	..
Steels			
C 0·08	Annealed	100–110	22–25
	Worked	160	31
Steel (mild)	Annealed	120	30
C 0·2–0·3	Worked	200	45–50
Steel (medium carbon)	Annealed	130	35
C 0·35	Worked	250	50
Steel (high carbon)	Annealed	140	35–40
C 0·5	Worked	300–400	60–80
Steel (stainless)	Quenched	250	..
18 Cr 8 Ni 0·1 C			
Tantalum	Annealed	60	22
	Worked	260	60–70
Thorium	..	..	35

TABLE VI (a) *continued*

Metal or alloy	Condition	B.H.N. (kg./mm.²)	Ultimate tensile strength (tons/in.²)
Tin 99·95	Cast, annealed	5	1·4
	,, rolled	6	1·5
Tin-babbitts			
Sn 91 Sb 4·5 Cu 4·5	Chill cast	17	4–5
Sn 89 Sb 7·5 Cu 3·5	,, ,,	24	5–6
Sn 83·4 Sb 8·3 Cu 8·3	,, ,,	30	5
Sn 65 Pb 18 Sb 15 Cu 2	,, ,,	23	..
Titanium, pure	Annealed	200	35
	Cold-worked	..	50
Tungsten, pure	Sintered	260	..
	Heavily swaged	490	..
	Heavily drawn wire	1,000	270
Tungsten carbide (Co 3–12)	Cemented	1250–1460	..
Wood's metal	Cast	16	4
Zinc, pure	Cast	30	..
	Rolled	35	8–12
Zinc, commercial (Pb 0·08)	Hot rolled	40	10
Zinc die-casting alloys			
Al 4 Mg 0·04	Cast	80	18
Al 4 Mg 0·04 Cu 1	Cast	90	20
Al 4 Mg 0·04 Cu 3	Cast	100	24

TABLE VI (b)

Physical Properties of Typical Indenter Materials

Material	Young's modulus 10^{12} dynes/cm.²	Hardness kg./mm.²
Steel (Ball bearing steel)	2	900
Tungsten carbide (sintered. 3–13% cobalt)	6·8–5·6	1,600–1,200*
Sapphire (synthetic)	3·7	1,600–2,000*
Silicon carbide	4	2,100*
Boron carbide (moulded)	5	2,230*
Diamond	Cubic axis 7·2 / Diag. of cube face 9·3 / Diag. of cube cell 10·2	6,000–6,500*

* Knoop hardness. See *Industrial Diamond Review*, 1940, **1**, 2; ibid. 1945, **5**, 103.

NAME INDEX

SUBJECT INDEX

Adhesion of metals, effect of released elastic stresses on, 90–3.

Area of contact, general, 141–53; real, 145–53.

Asperities, deformation of, spherical, 61–4, 141–3; conical, 143–5.
   role of, in determining area of real contact, 141 et seq.

Brinell hardness, 3, 6–17.
   comparison of, with Meyer's hardness, 11–12, 164–5; with Monotron hardness, 111, 164–5; with Rockwell hardness, 111, 164–5; with Vicker's hardness, 58–9, 99, 106, 112–13, 164–5.
   definition of, 6, 155.
   effect of surface roughness on, 14, 63–4.
   of very hard metals, 55–60.
   processes involved in, 93–4.
   strainless indentation in, 15–16, 65.
   table of values, 167.
   ultimate tensile strength and, 16–17, 79–83, 166.
— indentations, elastic recovery of, 14, 85–94, 118–20; piling-up and sinking-in of, 15; shallowing of, 14, 85–94; yield stress in free surface of, 72.

Coefficient of restitution, 129–30.

Conical indenters, indentation by, 95–114.

Deformation of metals; see elastic properties and plastic deformation.
— by spherical indenters, elastic, 44–6, 85–94, 126–9, 130–1.
   plastic, complete plasticity in, 47–9, 51; Ishlinsky solution of, 48–50; onset of plasticity in, 46–7.

Diamond indenters, deformation of, 58.
   hardness of, 56, 101, 171.
   use of, 3, 55–9, 98, 100.

Dynamic hardness, 4, 115–40.
   comparison of, with static, 137–8.
   definition of, 115.

four main stages involved in, 116–17.
   meaning of, 138–9.

Elastic equations; see Hertz's equations.
— limit; see yield stress.
— properties of indenters, effect of, on Brinell hardness, 55–60.
— — of metals, comparison with plastic, 25–6.
   effect of, on elastic recovery, 85–94, 118–20; on real area of contact, 147–8; on onset of plastic deformation, 53–4, 61–2; on rebound, 118–24, 126–8; on time of impact, 130–1.
— recovery of indentations, 14, 85–94, 118–20.
   effect of, on adhesion, 90–3; on area of real contact, 147–8.

Electrical conductance, use of, in determining time of impact, 133–7; in measuring area of real contact, 148–53.

Firth hardness, 98.

Friction, effect of, on indentation by conical indenters, 95–7; by flat punch, 37–40; by flat circular punch, 41–2; by spherical indenters, 49.
   on plastic deformation, 33, 37–40, 41–2.

Geometric similarity, principle of, in Brinell hardness measurements, 6–11, 67–9; in Vickers hardness measurements, 101, 105.

Hardness, conversion table, 164–5.
— definition of, 1, 6–7, 95–6, 115–16; meaning of, 93–4, 112–13, 138–9; typical values of, 167–71; see under Brinell, Dynamic, Firth, Knoop, Ludwik, Monotron, Pendulum, Rockwell, Scratch, Shore, Vickers.

Hertz's equations, for elastic deformation, 44–6; for elastic recovery, 86–8,